earth user's guide to

Permaculture

second edition

WOW!
LET ME
IN

Rosemary Morrow
Illustrated by Rob Allsop

Kangaroo Press

'May the Earth always speak to your spirit.'

EARTH USER'S GUIDE TO PERMACULTURE
This Second Edition published in 2006
First published in Australia in 1993 by Kangaroo Press
An imprint of Simon & Schuster (Australia) Pty Limited
Suite 2, Lower Ground Floor
14–16 Suakin Street
Pymble NSW 2073

A CBS Company
Sydney New York London Toronto

Visit our website at www.simonsaysaustralia.com

Distributed in the UK and Europe by Permaculture Magazine
www.permaculture.co.uk

Cataloguing-in-Publication data:
        Morrow, Rosemary.
        Earth user's guide to permaculture.

        2nd ed.
        Bibliography.
        Includes index.
        ISBN 0 7318 1271 9.

        1. Permaculture. 2. Organic gardening. 3. Urban
        agriculture. I. Allsop, Rob. II. Title.

        631.58

Cover design by Caroline Verity Design
Internal design by Avril Makula
Typeset in 10.5pt on 14.5pt Memento
Printed in China through Phoenix Offset

10 9 8 7 6 5 4 3 2 1

Foreword

Permaculture has its origins and strongest foundations in Australia, but it is also a global movement with projects and groups in hundreds of countries. Its inclusion in *The Macquarie Dictionary* is an indication of its status as almost a household word. In Australia permaculture has provided an ethical and environmental design framework for personal and household self-reliance, rural resettlement and community development. But the total scope and scale of permaculture in Australia is dwarfed by its growth in Europe, North America and Japan. In affluent countries permaculture has provided the environmentally aware with ways to reduce their ecological footprint through a greater degree of self-reliance. It has also spread to many less affluent countries of Asia, Latin America, Africa and Oceania through development projects. In this context, permaculture has been potent in the provision of basic needs using local resources.

Rosemary Morrow's decades of teaching and activism span the local and the global aspects of permaculture in rich and poor countries. *The Earth User's Guide to Permaculture* is very much a manual of practical permaculture. It is an especially useful book for home and land owners in Australian temperate regions but is also applicable more widely. This revised and updated edition shows a strong ethical sense of the global context as well as practical experience and examples of how the same concepts apply in very different climates and cultures.

Permaculture is sometimes understood as the work of one man, Bill Mollison. In fact, a diversity of people have contributed to its foundations, growth and spread, including Rosemary Morrow. With this new edition of *Earth User's Guide to Permaculture*, Rosemary Morrow consolidates her substantial and ongoing contribution to the understanding and application of permaculture as a design system for living lightly and yet abundantly on this earth.

DAVID HOLMGREN
Co-originator of the permaculture concept

old-growth and natural forests

our atmosphere

wildlife and endangered species

heritage food plants

care for the earth

care for people

soil life

redistribute surplus

all freshwater and marine systems

all people and cultures

rare breeds farm stock

The three attitudinal principles of permaculture:
care for people, care for the Earth, and redistribute surplus to one's needs.

Contents

Acknowledgments

‘You can’t do anything on your own.’ (Vietnamese saying)

Thank you, Rob Allsop, for your careful and painstaking work and wonderful friendship. Thanks also to Liz Connor, Sue Girard, Jill Finnane, Lyndall Sullivan, Veechi Curtis, Leanne Huxley and her family for the water ‘policing’ statistics (Leanne’s word), Brian Coates, and Lyn Godfree for stretching my ideas and adding their experiences. My friend Margot Turner had the inspiration and added to the chapter on permaculture at work. She also supplied the materials for the chapter. Her support has added considerably to the scope and relevance of permaculture at work.

The motivation for this book comes from the hundreds of students who have completed short and long courses with me, and whose evaluation always asked for ‘more tangibles please’. I truly hope that this book will go some way towards meeting their needs.

There have been people in the Blue Mountains where I live who have carried me through computer despair when I seized up; in particular, Ted Markstein and Kai Green.

Thanks to Julia Collingwood at Simon & Schuster and Katrina Herborn, my doctor, who convinced me to work on this new edition after I had a stroke. And to Glenda Downing, my editor, who has become a friend through lots of sympathetic discussions.

Thanks are due also to permaculture teachers with whom I spent some hours in discussion.

And how can I say thank you to people in other countries, where I brought the ethics and principles and they added the rest … Gia, Nguyen Van Gia, Le Phuong, Em Ponna, Kosal Neary, Koli Stafa, Edi Mullah and many, many others who started permaculture, never having seen it in the field. You are wonderful.

Introduction

The world has changed. Most of Earth's citizens are now aware of the degradation of water, soils, forests and air quality, and that this destruction is accelerating. We know we are living in an age of energy and freshwater peak and decline and the future does not look pleasant unless we move creatively into the future, changing our attitudes and lifestyles. However, it is not simply enough to slow down the destruction; we must work to reverse it. Permaculture provides us with the skills and knowledge to do this in ways that are creative and satisfying. Much of permaculture is repair.

We appear to have reached a 'tipping point' or critical mass where it will now only take a small change for systems to break down. We can think of it as putting more and more weight on a stick until it suddenly breaks. But all of us can easily reduce our impact on Earth's scarce resources by acquiring knowledge and skills. We can guard our future by living lightly and consuming thoughtfully.

Not only have our ecosystems become more vulnerable but our lives are now also directly threatened. In this edition attention is given to individual and community resilience and self-reliance through design and community structures. Permaculture may be the only integrated wholistic model we have to build a more secure future or, at least, help us cope with an uncertain one.

Some good has come from awareness of our human impact. Many of us know the value of insulation, water harvesting and native vegetation. Compost and mulch are no longer mysteries. Permaculture is a recognised word. More people are rejecting plastics and many more are reducing waste. Some people use ethical investment, others are aware of their ecological footprint. While this is

not enough to change the trend towards a diminished and difficult future, it provides hope that we can create pockets of stability and regeneration. Governments are not to be depended on for the leadership we require. Permaculture is at the vanguard of change through individual action, advocacy and co-operative groups.

This book is a practical and simple guide, a textbook for developing permaculture designs for your home, land and life. Fundamentally, permaculture is design science and in this edition design is emphasised. There are clear design aims to work from as you create strong perennial landscapes, work practices and sustainable buildings. The book covers most of the syllabus of the permaculture design course issued by the Permaculture Institute as well as additional ideas that the world has given us in the last decade.

Because other books such as David Holmgren's welcome and timely addition to the canon, *Permaculture: Principles and Pathways Beyond Sustainability*, and Bill Mollison's 'bible', *Permaculture: A Designer's Manual*, already comprehensively cover the theory of permaculture, this book is less about theory and more about doing.

It is suitable for you if you are:
- striving to live more sustainably, whether you have land or not
- interested in ecologically sustainable development
- a student of geography, agriculture, landscape design, horticulture or architecture
- interested in permaculture yet don't want to do a face-to-face course
- extending your knowledge of permaculture.

If you are a beginner you will be able to effectively transform your land and life, and if you are an experienced permaculture practitioner, then you can try some new ideas and move towards greater self-reliance and sustainable living.

This book will be most beneficial if you apply it to the space where you live and work. It does not matter whether you live in a small city apartment with a windowbox, in a house and garden in the suburbs, or on a farm in the country. The same principles apply for becoming more sustainable and resilient.

I have chosen Rob's place, a small suburban home in the Blue Mountains, west of Sydney, and Rosie's farm, 32.5 hectares located to the west of the Blue Mountains, as examples with differences in climate and scale resulting in different design strategies. They will demonstrate the flexibility inherent in working from permaculture principles.

Glance through the book before you start so that you gain an understanding of the whole. It is divided into five main parts. The first two parts are mainly background information, applied science and skills you will need to enable you to analyse what is happening on your land. The next three parts provide you with information on how to design a site and then implement that design.

Four icons appear throughout the book:

 The *globe* indicates our ethical task to sustain the Earth.

 The *foot stepping lightly* indicates your design objectives.

 The *heavy foot* indicates a lack of design objectives and signifies unthinking destructive action.

 The *garden fork* appearing at the end of each chapter beside 'Try these' asks you to try something. These activities will give you the experience you need to feel confident as you move towards your final integrated design.

You will also need a notebook to record your research, observations and planning.

Start now and let your life be enriched. Be part of the solution and remember the future.

PART ONE

An observing and appraising eye

Permaculture is the science of applied ecological design and, as with all sciences, you need some knowledge and skills. In permaculture, the fundamentals are water, soils, climate and plants. It is essential for good design work that you are a critical observer of all these elements. Every site is different and the rigour and care you bring to your observations will make all the difference between a good sustainable design and a poor one.

Like Professor Julius Sumner Miller, the physicist who popularised science on television, you should ask often, 'Why is it so?' And, 'Is it particular to my site or is it universal?' In design we proceed from the general to the particular, and your skills in questioning, observing and deducing will improve as you practise them and your knowledge and experience grow.

Underlying all design work is the concept that everything lives and exists within patterns of time, birth, growth and decay. The better you perceive and work with patterns, interlocking and overlaying, the better your designs.

And trust yourself, your observations and your intuition, which will subliminally inform your work.

CHAPTER 1
Starting permaculture

Before you start permaculture, it is useful to know something of the history and foundations upon which it is built. Permaculture was developed by Australians Bill Mollison and David Holmgren in the 1970s in response to Earth's soil, water and air pollution by industrial and agricultural systems; loss of plant and animal species; reduction of natural non-renewable resources; and destructive economic systems.

Bill and David reassembled old wisdom, skills and knowledge of plant, animal and social systems, and added new scientific knowledge. Permaculture was born. The result was a new way of sustaining and enriching life without environmental and social degradation. Although many of the parts of permaculture were familiar, it was the overall interlocking framework and pattern that was new, exciting and different.

Unlike modern science, which is reductionist, reducing everything to its smallest components, permaculture places itself squarely on the shoulders of ecology, the study of inter-relationships and the interdependence of living things and their environment. Permaculture offers an understanding of how biological processes are integrated, and it deals primarily with tangibles: plants, soils, water, animal systems, wildlife, bush regeneration, biotechnology, agriculture, forestry, architecture, and society in the areas of economics, land access, bioregions and incomes tied to right livelihood.

How do you make sense of these and weave them into a design? Your tools are observation, analysis and synthesis, and the result is applied design for sustainable living.

There are many definitions of permaculture. Here is one: 'Permaculture is about designing sustainable human settlements through ecology and design. It is a philosophy and an approach to land use which weaves together microclimates, annual and perennial plants, animals, soils, water management and human needs into intricately connected productive communities' (Bill Mollison and Reny Mia Slay, *Introduction to Permaculture*).

And, as Mollison goes on to say, 'Permaculture is ... working with nature rather than against nature ... of looking at systems in all their functions rather than asking only one yield of them; and of allowing systems to demonstrate their own evolutions.'

The main features of the permaculture approach are summarised as follows:

- It is a synthesis of traditional knowledge and modern science applicable in both urban and rural situations.
- It works with nature and takes natural systems as models to design sustainable environments that provide for basic human needs and the social and economic infrastructures that support them.
- It encourages us, and gives us the capacity and opportunity to become a conscious part of the solutions to the many problems that face us locally and globally.

Why practise permaculture?

For you permaculture can be the creative alternative that our society is not offering. It opens doors to a simpler, better-quality life and it is empowering because anyone can 'do' it. There are no barriers of age, sex, religion, education or culture.

You start small, so the scale and costs are not prohibitive, and whether you have 1 square metre or 1 million hectares, permaculture will work. The act of working closely with the earth and taking responsibility for how you treat it gives you greater and greater rewards as you respond to the positive environmental changes. If, over your lifetime, you build or retrofit a simple, non-polluting home and grow your own food, build soil, and care for natural vegetation, then you will have lived a full, creative and interesting life with great personal freedom, satisfaction and autonomy.

To become a permaculture designer, you will need to complete four main tasks as you work through this book:

1. Carry out an inventory of your house and land.
2. Make a site analysis of your land, showing its strengths and weaknesses in detail.
3. Make your final step-by-step design for the land to achieve permaculture goals.
4. Record your observations regularly in your observation journal.

Understanding patterns

Pattern language is basic to the study of permaculture. All life exists within patterns of birth, growth and decay. Life patterns are in motion and each element has its own lifecycle, from minutes to thousands of years. Stones, wind, sun, seas, stars and rivers also exist within patterns. Patterns are embedded within other patterns; patterns overlap in time and space.

Some patterns are more prevalent than others. Because of patterns, life perpetuates itself and exists sustainably; for example, the pattern of ocean currents bringing changes of seasons, the patterns of fruit ripening and seed scatter. Nature is a repetitive series of time and space patterns. You can learn to read these patterns and understand how they work by observing your land.

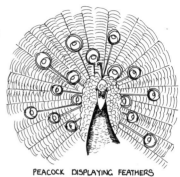
PEACOCK DISPLAYING FEATHERS

When you notice a pattern, watch how it functions and evolves. For example, after clearing land, weeds emerge to protect and change bare soil.

All patterns have mathematical relationships. Animals build nests and hives, rivers flow and trees branch according to mathematical relationships— the honeycomb has hexagonal cells, for example. Nature has several main patterns: circles, spirals and networks, and all of these have mathematical interpretations. Interestingly, architecture and music, like patterns in nature, are also mathematically underpinned. Often the mathematical relationships are expressed in ratios. Elements of patterns which lead to good design are:

- symmetrical, and asymmetrical, balance in shapes and patterns
- proportion (ratios) in nature
- play of patterns on patterns and patterns within patterns
- repetition of patterns
- perception and integration of existing patterns.

All patterns in nature have a time component. The patterns are a sequence of events which are repetitive and often predictable. Some examples are the annual movement of the earth around the sun, seasonal rainfall, seed dispersal, and the migration of animals. Figure 1.1 shows the sun movement and shadow pattern in a day.

Patterns can be dimensional, such as lines of weeds along roads, swales along contour lines, river flow down valleys, the deposition of alluvial soil

SNOWFLAKE

Figure 1.1 The sun movement and shadow patterns in a day.

Applications of pattern knowledge

You can use your and others' knowledge of patterns to increase productivity, protect your designs, provide a buffer against disasters and reduce the use of resources.

Once you have made an observation then correlate it with time; for example, seedpods shatter along riverbanks, the seasonal emergence of seedlings, or the growth of a forest. They are often triggered by factors such as temperature, rainfall, day length, altitude, soil acidity or alkalinity. Often the 'trigger' has a time pattern and correlates with some other natural variable. For example, soil temperature and seed germination, or day length and flowering.

 ## Our ethical task is to:

- appreciate patterns
- achieve a harmonious resolution of problems
- increase and protect resources
- achieve specific design outcomes.

 ## Our permaculture design aims are to:

- work with patterns that already exist
- impose patterns to achieve specific results
- understand and use edge effects (where two ecosystems or microclimates meet)
- manipulate the flow of air and water
- understand time as a dimension of patterns.

 ## If we don't have design aims:

- there will be many problems with scale and proportion
- possible environmental destruction or degradation, for example seed infestation, will result.

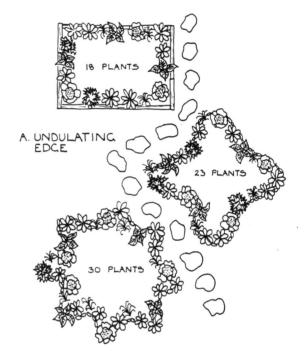

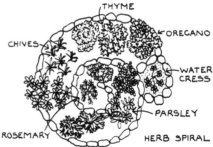

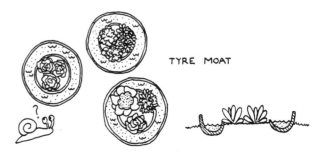

Figure 1.2 Small-scale design techniques.

on a very hot day. You will design sites more effectively, place elements more accurately and work more closely with nature by paying attention to your local time cycles and patterns and working within them.

Most traditional strategies and techniques develop as a result of understanding local patterns. The diagrams in Figure 1.2 show some of the most common patterns used in permaculture designs.

There are certain shapes or patterns that function extremely well. One used repeatedly in permaculture is a wavy or undulating edge. This increases the length of edge compared with a straight line. It is used in designing paths and keyhole planting beds.

Another is the circle, which is a strong pattern because it has a small edge in comparison to the inside area. If you wish to conserve resources, as people do in dry areas, then circle beds are very efficient in using resources and provide good protection for plants.

A spiral is appropriate where you require several microclimates and vertical growth to conserve space, or in regenerating bushland where you need plants to protect each other while they become established. When you have to choose a size, you apply patterns in order to choose correctly. For example, how often have you heard of people putting huge forest trees in small gardens or big dams in deserts or small shrubs in a large windy landscape, or getting the spacing wrong when planting seedlings? These mistakes are often a result of bad design, such as a huge garden where there is no water or no labour to maintain it, or too small a farm to work efficiently while still protecting the water and soil.

Try these:

One of the key skills you will need to develop to design permaculture sites is the ability to observe. Observation is also important in defining problems, recognising the potential and patterns of a site, identifying their origins and 'getting it right'. Always look for relationships among patterns. Get an observation journal and some coloured pens to record what you see. You may like to add drawings or photographs. You can also model observations in sand, clay or plasticine.

1. Write the season, date and the year.
2. Now observe: walk outside or look out your windows and record what you see. Your entries may look like this:
   - New red leaf tips as a halo over the eucalypts.
   - Two white-backed currawongs in the eucalypts.
   - Heavy clouds coming in from the west.
   - New growth on the heather is above the old flowers.
   - Roadside grasses are in flower.
   - New growth on wattles.
   - *Prostanthera* (mint bush) has just finished flowering.
   - The first silvereyes have arrived and are perched in the cotoneaster.

   Now go a little further with your observations and use your senses:
   - What do you hear?
   - What can you smell?
   - What can you feel?
3. Correlate your observations; for example, 'It's a dry hot day *and* there's an increase in red spider mite.'
4. Repeat this exercise every day and become aware of patterns.
5. Look out for resources you could need later to implement your design. These resources are often someone else's rubbish, such as old bricks, timber, straw, grass clippings, or old fruit tree varieties you may want for grafting. Note their location, quality and possible uses.

CHAPTER 2

Permaculture's ethical underpinnings

When permaculture started there wasn't the public debate we now have for the need of ethics in public and private life. And yet Bill Mollison and David Holmgren founded the practices and concepts of permaculture on ethics and principles.

Ethics provide a guiding sense of obligation and are the broad moral values or codes of behaviour against which ideas and strategies can be tested. Think about this: do soil scientists benefit the earth? Consider chemical fertilisers and pesticides. What would happen if scientists had ethics and not simply research goals?

Below are the permaculture ethics. They are not in order of priority; all apply equally.

- Care for people.
- Care for the Earth.
- Redistribute surplus to one's needs (seeds, money, land, etc.).

Principles have narrower applications because they are more like a specific set of guidelines. Principles define the limits of ethics. It is fairly easy for everyone to agree on ethics but there will be different opinions about principles, which may change over time. Bill Mollison presented one early set of principles and then a later one, and recently David Holmgren has given another set. Some of them overlap.

I have arranged the permaculture principles in a different way because of the increasing urgency to repair and regenerate Earth's natural systems. They are divided into three groups, which reflect what I believe to be the priorities when starting on effective restoration of landscapes. These three groups are:

- design principles
- strategic principles
- attitudinal principles.

In Tables 2.1 and 2.2, the design principles and strategic principles have been arranged in priority for action. If you wish, you may reorganise them. The best way to use these principles is to turn them into questions for yourself. For example:

- Have I considered water, energy and biodiversity sustainability in many ways in my design?
- Am I working with nature?
- Have I started small and got it right?
- Have I started with water and energy?

Design principles

The design principles are based on the great ecological themes, such as water, energy and soil, which are essential for sustaining life processes.

ATTITUDINAL PRINCIPLES	
BASIC POSITION	OUTCOMES
• WORK WITH NATURE NOT AGAINST IT	• RESULTS IN MINIMUM NEGATIVE IMPACT AND LONG TERM SUSTAINABILITY
• VALUE EDGES AND MARGINAL AND SMALL	• SMALL AND DIFFERENT CAN BE VITAL
• SEE SOLUTIONS INHERENT IN PROBLEMS	• OVERCOMES BLOCKAGES TO DESIGN AND IMPLEMENTATION
• PRODUCE NO WASTE	• MOVE TOWARDS A CLOSED ECOSYSTEM
• VALUE PEOPLE AND THEIR SKILLS AND WORK	• DRAWS PEOPLE IN, ENABLES, APPRECIATES AND SUPPORTS THEM
• RESPECT FOR ALL LIFE	• THE DELIGHTS OF ALL NATURAL AND CULTURAL DIVERSITY ARE VALUED
• USE PUBLIC TRANSPORT AND RENEWABLE FUELS • CALCULATE 'FOOD MILES'	• MOVE TOWARDS PEOPLE - SCALED, SUSTAINABLE URBAN PLANNING, FRIENDLIER PLACES AND LESS POLLUTION • SUPPORT LOCAL FARMERS, BIOREGIONAL PRODUCE, LOWER FOOD COSTS, TRUCK - FREE ROADS
• REDUCE YOUR ECOLOGICAL FOOTPRINT	• ACCEPT RESPONSIBILITY, SIMPLIFY YOUR LIFE, BECOME MORE SELF - RELIANT • REMEMBER THE FUTURE AND SAVE RESOURCES

Figure 2.1 Attitudinal principles of permaculture.

Strategic principles

These minimise negative environmental impacts caused by design errors and ensure that essential elements of natural systems are safeguarded and resources used appropriately They help you to get most of it 'right' the first time instead of having to correct mistakes later.

Attitudinal principles

The final category of principles broadens your perspective in analysis, design and implementation.

Attitudinal principles raise your awareness of the value of different elements affected by your design. They refine and clarify your design work (see Figure 2.1).

Permaculture in practice

Permaculture is not an armchair study. It is about acting and changing behaviours and skills on land and in society. Of course it is hard to imagine what a permaculture world would look like simply from ethics and principles. Figure 2.2 is a word picture of a permaculture world. From these, can you imagine

TABLE 2.1: DESIGN PRINCIPLES

Design principle	Where it applies
Preserve, regenerate and extend all natural and traditional permanent landscapes.	Watersheds, valleys, roadsides, remnant forests, ridges and steep slopes and your backyard.
Water: conserve and increase all sources and supplies of water, and maintain and ensure water purity.	Catchments, tributaries, soaks, wetlands, rivers, lakes, aquifers, springs and estuaries, including traditional water-supply systems, for example. Balinese terraced rice fields and Middle Eastern underground tubes and melt-water canals, and your backyard. Care for the health of marine systems.
Energy: catch and store energy by all non-polluting and renewable means.	Look to sun, wind, wave and geo-thermal sources. Catch by vegetation. Convert broad-scale monocultures to permanent diverse systems, water bodies, and biomass (plants and animals) of all types. Use passive solar design and technology. Use wind and wave energy for power generation.
Biodiversity: preserve and increase biodiversity of all types.	From rainforest to deserts. From invisible to macro. In niches, habitats, seeds, pests, human settlements, religions, knowledge, skills and attitudes.

what a permaculture farm, balcony, community or landscape would look like?

Permaculture makes use of many strategies and techniques to help with the principles. Strategies are techniques implemented over time and tell us 'how and when' to do something. Techniques just tell us 'how' to do something. Although all permaculturists subscribe to the same ethics and principles, their strategies and techniques are often very different because no two environments are the same. Only imagination limits the development of new strategies and techniques to support the ethics and principles.

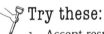

Try these:

1. Accept responsibility for how you live. Simplify your life, become more self-reliant, remember the future and save resources.
2. Write in your observation journal the principles that seem most important to you and how their application would affect your life. Write another list of those that really appeal to you and the reasons why. And those that challenge you and why.
3. Observe your place very closely and compare it to the table of characteristics of a permaculture

TABLE 2.2: STRATEGIC PRINCIPLES

Strategic principle	Where it applies
Focus on long-term sustainability.	Careful thinking.
Co-operate don't compete.	Share best knowledge and practice.
Design from patterns to details.	See the whole picture first.
Start small and learn from change.	Avoids expensive errors.
Make the least change for the largest result.	Efficient and economical detail.
Make a priority of renewable resources and services.	Establishes a feedback loop to long-term sustainability.
Bring food production back to cities.	Empowers food security and risk avoidance.

CHARACTERISTICS	APPLICATIONS
SMALL SCALE LANDUSE PATTERNS	MOST MARGINAL LAND RETURNED TO NATURAL ECOSYSTEMS TO COLLECT AND RETAIN WATER, SOIL AND INDIGENOUS SPECIES. LANDSCAPES ARE VARIED AND INTERESTING
INTENSIVE RATHER THAN EXTENSIVE	MOST WORK CLOSE TO HOME, EASY TO MANAGE, HUMANE PLANT AND ANIMAL SYSTEMS, PERHAPS BRING SHEEP AND GOAT (NATIONAL HERDS) TO CITIES IN HOME GARDENS
DIVERSITY WITHIN HABITATS	DIVERSITY OF SPECIES, CULTIVARS, YIELDS, NICHES, FUNCTIONS, SOCIAL ROLES, WORK AND CHOICES
INTEGRATION OF MANY DISCIPLINES	AGRICULTURE, AQUACULTURE, FORESTRY, ANIMAL HUSBANDRY, WILDERNESS AND SOCIAL BEHAVIOUR (ECONOMIC, RELIGIOUS ETC.)
USE OF WILD AND DOMESTIC SPECIES	POSSIBILITY OF INNOVATIVE USE OF RABBITS, KANGAROOS, GUINEA PIGS, SNAKES, PIGEONS AND DOGS
LONG TERM SUSTAINABILITY	AIM FOR PERPETUATION OF SYSTEMS WHICH CAN ADJUST TO CATASTROPHES SUCH AS THERMAL POLLUTION AND LOSS OF OZONE, WITHOUT LOSS OF SPECIES RICHNESS
USE OF NATURALLY INHERENT TRAITS OF LAND, PLANTS AND ANIMALS	ENERGY, WATER AND SOIL RESOURCES ARE CONSERVED, REBUILT, SELF-REGULATED AND SELF-REPAIRING

A TREED LANDSCAPE	CULTIVATION IN CLEARINGS PROTECTED BY PERENNIAL PLANTING AND NATURAL FORESTS
WHOLE SITE PLAN FOR WATER SECURITY	DROUGHT AND FLOOD PROTECTED BY DAMS, TANKS, VEGETATED CREEKS, STREAMS, RIVERS AND WETLANDS, RECYCLING AND RE-USE OF WATER

Figure 2.2 Characteristics of a permaculture landscape.

system. Give yourself ticks for good approximation.

4. List as many different plants and animals in your garden as you can, and make sure you find out the names of those you don't know. This is a measure of the present diversity there. Later you will see how this can be increased.

5. Using the table of characteristics, visualise your ideal permaculture landscape then realise it by writing a poem, making a collage, constructing a model, painting or sketching—whatever medium suits you.

6. From now on, support local farms, truck-free roads, locally grown (that is, bioregional) and organic food.

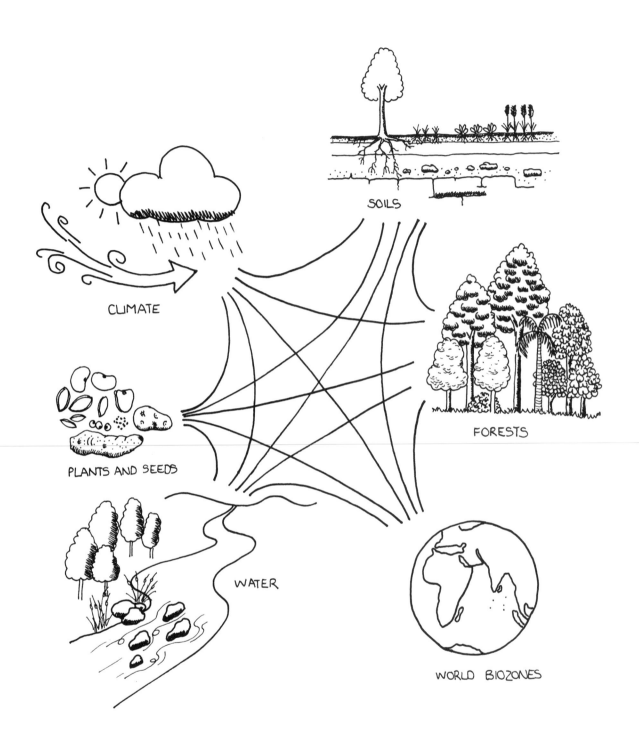

SOILS

CLIMATE

PLANTS AND SEEDS

FORESTS

WATER

WORLD BIOZONES

PART TWO

Ecological themes in permaculture

You have just explored some basic permaculture ideas. Most sciences, including modern agriculture and horticulture, have developed through a reductionist approach; that is, by taking everything apart—the soil, plants and insects—and examining each in isolation and greater and greater detail. Rarely are the parts fully put together again. This means that the effects of connections and surrounding influences on what is being studied are ignored.

In permaculture we are very interested in the connections between things and the influences on them. However, in this part of the book we will start by taking life apart to look carefully at its disciplines before putting them together to create something greater than the parts.

In Part Two we will look at major design themes which you will use across a whole site so that you learn to develop a 'whole site water plan' or a 'whole site soil management plan' or a planting plan. Simultaneously, you will be building up a comprehensive site analysis of the land you wish to design. In this part we will also look at:

- biozones and traditional land use in other parts of the world
- the effects of inappropriate land use
- the reasons why traditional cultures are breaking down.

But first, we need to start with some ecological ideas.

CHAPTER 3

Ecology: life's networks

Ecology is the study of natural systems, their interdependence and inter-relationships. An ecosystem is a community of organisms interacting with each other and with their physical environment and functioning together as a complex self-sustaining natural system.

The study of ecology and cosmology over the last 30 years or so has introduced the idea of 'nested ecosystems', where separate microclimates are nested within a bioregion, nested within Earth's biosphere, nested within our solar system, nested within our galaxy, etc. In other words, local systems are autonomous in some senses but also interdependent with other systems in ways that we are only now beginning to understand. And at every level complexity is inherent.

Ecologists believe in what's known as 'the ecological imperative', which says that humans are part of ecosystems and must acknowledge their inter-relationship with and interdependence upon them. Permaculture is often called 'the cultivated ecology' because of its goal to integrate and transform human societies so they live in sustainably designed and highly productive ecosystems where self-interest is aligned with the common good.

 ## Our ethical task is to:

- design ecosystems that maximise the number of productive species
- use energy and matter effectively
- move towards ecosystem perpetuation.

 ## Our ecological design aims are to:

- preserve genetic diversity
- respect the right to life of all species to contribute to ecosystem structure
- allow ecosystems to evolve under changing conditions—the land forgives us
- use species and habitats sustainably so the essential life-sustaining processes can continue intact
- design closed systems in which all needs are met.

 ## If we don't have ecological design aims:

- closed systems are broken open
- artificial industrial systems are created which depend on non-renewable energy sources. For example, when industrial agriculture, called monoculture, removes biodiversity the ecosystems collapse. Forest removal is especially damaging and its destructive effects multiply because when the closed system affects other connected systems they also collapse—a bit like taking the bottom out of a

pyramid. There are serious flow-on effects in time and space. Some examples are:
- We break fundamental and often unknown laws of nature.
- Rivers loaded with farm chemicals choke and die.
- Clear-felled forests become eroded and disturb rainfall patterns.
- Drained wetlands lose habitat for migrating birds.
- The atmosphere, overloaded with gases, causes global warming.

Closed systems

Natural ecosystems are known as closed systems because they meet their own needs internally. They supply pest management, nutrients for all species, temperature control, soil building and maintenance, wind control, pollination, germination and pruning. They don't need any human management or inputs. A well-designed mature permaculture system approximates a closed system in meeting as many of its needs as possible.

Look how the Banana Circle in Figure.3.1 is an example of a designed cultivated ecosystem.
- The acacia provides windbreak protection for the bananas, seed for the chickens, mulch for the ground and puts excess nitrogen into the soil. It needs weed control and nutrients.
- The growing banana plant uses surplus nutrients and holds soil against erosion. The stem holds water. It requires windbreaks and pest control.
- Chickens eat any pests of the bananas (a protein source), obtain water from the base of the banana, supply nutrients in the form of manure and eat the seeds of the acacia. They need water, seed and protection.

The functions of healthy ecosystems are to:
- create and support life
- clean air, clean water, and sometimes toxins through various filters
- regulate the atmosphere through recycling carbon and nitrogen
- build soil together with soil micro-organisms.
- manage pests and diseases

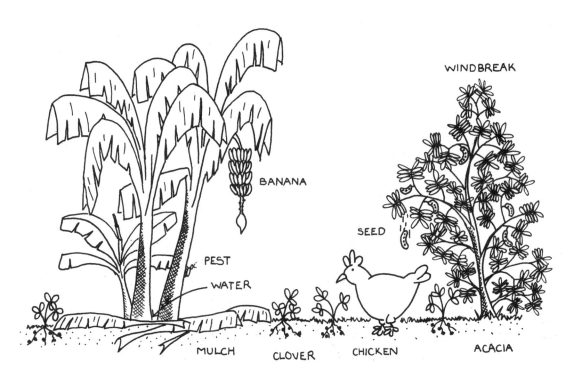

Figure 3.1 A cultivated ecosystem. Ecosystems can be consciously designed to be productive and sustainable.

- perpetuate themselves
- create highly integrated structures—often finely tuned
- become closed systems.

These concepts of natural functions translate into practices which you will learn. So, for example, biological water cleansing requires a design that approximates a wetland's because wetlands cleanse water.

Gaia

The Gaia theory was developed by the nuclear scientist James Lovelock in the 1980s. He hypothesised that the Earth could be imagined as one entire living super-organism evolving over the vast span of geological time. He believed that the Earth is a complete organism and, like all others, is essentially self-regulatory and keeps itself in good health. In his theory he suggested there are various organs which are critical to the survival and health of Earth. He made the analogy that forests are like the kidneys filtering and cleaning water; oceans are like the lungs; rivers are like the blood system; rocks are like the bones; and that when organs are destroyed, so the life of the organism is severely threatened. Each organ is intimately related to all others and disease in one affects the others. So, for example, destruction of forests causes imbalance in the Earth's atmosphere and hydrology.

Organs can only take so much destruction before a critical point is reached and they collapse. Lovelock suggests that when forest cover falls below 30 per cent of the Earth's surface then other systems will collapse. In permaculture we set goals to clean water, protect rivers, and maintain or create at least a 40 per cent permanent tree cover.

He also suggested that human impact on the earth should be seen in the same light as a virus's impact on an organism. The theory is now widely accepted. If it is proven correct we have a very unhappy future ahead of us. So now we must audit, monitor and regulate our consumption of non-renewable resources. We have two principles to guide us.

- *The precautionary principle*, which states that we should take seriously and act on any serious or destructive diagnosis unless it is proven erroneous. This applies to the Gaia theory. It is a basic principle of ethical design.
- *The intergenerational equity principle*, which states that all future generations have the same rights as us to food, clean water, air and resources.

Our ecological footprint

Since writing the first edition of this book, the exciting and useful concept of a measurable ecological footprint has emerged as a way for everyone to know, understand and compare their impact (footprint) on Earth and its resources.

The ecological footprint is a measure of a person, town, city, or nation's use of resources. It measures use of energy, water, food, clothing, housing materials and transport. Each of these is calculated and given in hectares. There is now no excuse for not knowing what we cost the Earth, and because our footprint is measurable we can also reduce it. To reduce our ecological footprint is one of the goals of permaculture design.

Anyone can measure their footprint. Schoolchildren now regularly calculate theirs. For a fair and equitable use of Earth's resources, each person requires a 2.5-hectare footprint to meet all their needs and sustain the processes of nature that make life possible. At the moment, Australians have a 7.6-hectare footprint. At this rate we require three Earths to continue to meet all our needs.

The larger your footprint, the more of the Earth's resources you use. For example, a child scavenging a dump in Mumbai has almost no footprint, whereas children in Amsterdam can have a footprint of 9 hectares. On average, the Dutch have a footprint of about 7 hectares. The world cannot afford this degree of consumption. It means that many nations send out and raid resources from all around the world, for oil, timber, heating, foodstuffs and clothing. A country like Afghanistan has a tiny footprint because it produces almost everything its population uses. The people build their own houses, grow most of their own food, use little transport and so do not contribute to reducing non-renewable resources. There is also little pollution or overloading of ecosystems. They are much nearer to being a closed system than most countries are.

Look at the website *www.myfootprint.com* and calculate your footprint. Then, after practising permaculture for a year, measure it again and you will find your footprint is reduced. You start by reducing your largest 'toe' or factor. If it is water then start there; if it is transport then reduce that.

The concept of 'food miles' is used for reducing your footprint for food. Food miles are a measure of how far your food travels and, consequently, of the resources used for it to get to you. We need to know the food miles of everything we eat.

A third important ecological measure is harder to find out but has been calculated for some products. This is the 'lifecycle cost', or the cradle-to-grave cost, of any product. Most often this is measured in terms of how much energy is required to:

- source the raw materials
- manufacture the product
- package, transport and market it
- dispose of it when discarded or useless.

This is usually referred to as the embedded energy cost. However, it can also be calculated in other ways. For example, every litre of bottled drinking water takes 200 litres of water to produce. This is its cradle-to-grave cost, or the real cost of finding water, piping it, cleaning it, making bottles, transporting them, packaging them and then tossing away the bottles and their effect on climate change. The cradle-to-grave cost of a pencil is 30 per cent of that of a ballpoint pen. A staple has a greater cost than a paperclip because the paperclip is reused many times. Every product has a cradle-to-grave cost. Any product with more packaging than an equivalent product has a greater cost. Always choose the product with the lower cost and boycott the other one.

The new science of networks

Network science is the newest science and was only defined in 2004. It explains the complexity and stability of ecosystems. Science until now has taught us about entropy—basically that things fall apart. According to this theory, the universe, left to its own devices, will degenerate and tend towards entropy, or ever-greater discord and randomness. And yet in permaculture we see things that do just the opposite, that seem to drive themselves to an ever-greater development of patterns and complexity, as happens with the evolution of ecosystems. In life we can go from a world in which everything is disconnected to a world in which everything is connected; for example, from putrid organic matter to composted humic acid.

According to the new science, integration and complexity rely on synchronicity and networks. Scientists have found that pairs and trio-links start first and these are called nodes. Then, as these make links, a giant component emerges in a very short time to form one huge 3D network called a cluster. So we go from a stage of disorder, of chaotic motion, to order where everything is linked. The point at which that happens is called a tipping point, or phase transition. When water turns to ice the tipping point is 0°C and all the molecules in water form ice. Nodes and clusters make this happen. Everything is a network: knowledge, epidemics, diseases, nutrient cycles, electricity grids, crickets chirping, nervous systems and roads.

Nodes and clusters dominate nature. The network structure is very efficient at passing information on and also enables us to understand why and how structures collapse. Take out a cluster with a high number of links and you get cascading failure—the systems collapses. This explains why, if you take out a keystone species such as bees, ants or some plants, a system will rapidly degrade.

So in a garden or on a farm, some species are nodes and others links. The daisy family, which many insects visit, is a node and bees are its links.

What will happen if nodes are destroyed or not included in a design plan? How do we make nodes and then link the nodes? We only know a little, but what we do know we use in our designs.

Permaculture has always known about nodes and links intuitively and has created links between species, especially keystone species. But now we have a name for it—network science—which gives us a clearer concept and the language to include it in our design thinking. We know that it is not the number of species as such that is important but the number of links between them.

How ecosystems work

Ecology is a young science and we really know very little about it. However, here are some starting concepts that are helpful for designers.

Energy flows through ecosystems

All life forms must have energy to drive their activities. The primary source of energy for life is light energy from the sun. Plants capture light energy through photosynthesis and turn it into chemical energy, such as carbohydrates, sugars, waxes and oils which are eaten by organisms and, in turn, supply them with energy (see Figure 3.2).

Figure 3.2 Plant a garden and watch the energy flow.

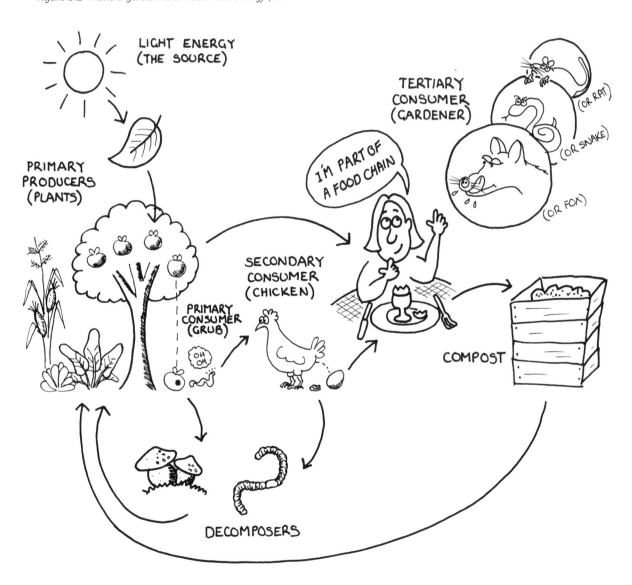

Energy moves through all living systems from the sun (the great power station in the sky) to the plants (primary producers), then to the herbivores (consumers), which eat seeds, grass, leaves or fruit and a variety of other organisms. They, in turn, are eaten by carnivores. Eventually, everything decays and ends up in the gut of the earthworm (decomposer) where the remaining energy is finally released by a bacterium as carbon dioxide and water.

Figure 3.2 shows the flow of energy through a system called a food chain. By growing plants, whether a vegetable garden, or a forest, you are initiating the capture of energy from the sun. It then flows through all organisms by a variety of routes, which form a web or network.

Energy can be lost from your system (when you take leaves and grass cuttings to the tip, for example), or you can save it and reuse it (by turning those cuttings into compost). When you are conscious of the flow through of energy you use it many times. When chickens eat your diseased fruit to make manure, which is fertiliser for your garden, you are using energy well.

Matter cycles

Matter consists of all the elements and molecules which make up the gases, vitamins, proteins, minerals and other nutrients of life. The total amount of matter in the world is constant. Yet each element can change to other forms. For example, iron may take one form in blood and another in rocks. As it changes form, it cycles through various organisms.

All matter cycles through living and non-living materials (air, rock, trees, animals, etc.) on Earth. The cycling of matter is driven by the sun and facilitated by the flow of energy. Cycles can be fast, as in the simplified nitrogen cycle shown in Figure 3.3, or slow as for uranium.

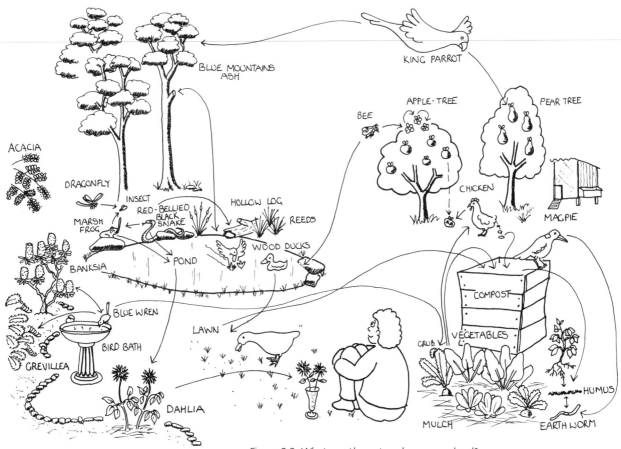

Figure 3.3 *What are the networks on your land?*

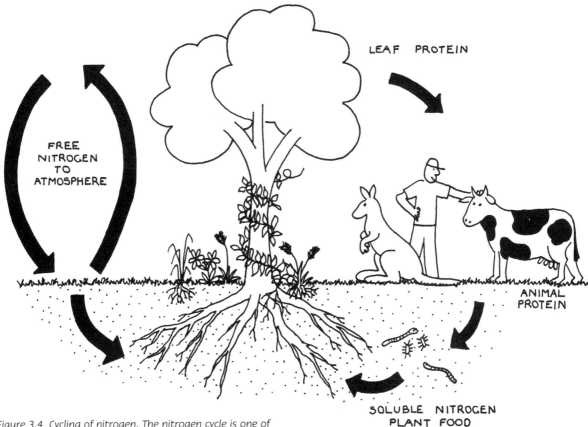

LEAF PROTEIN

FREE NITROGEN TO ATMOSPHERE

ANIMAL PROTEIN

SOLUBLE NITROGEN PLANT FOOD

Figure 3.4 *Cycling of nitrogen. The nitrogen cycle is one of the essential nutrient cycles which occur in ecosystems.*

Surplus causes pollution

When there is surplus to any system's needs it causes pollution, whether it's unburnt gases, chemicals in water, exotic animal invasion or loud noises. Pollution is human design failure.

When inputs are not fully utilised in cycles of matter it is called bioaccumulation; this is one type of pollution. Surplus phosphate fertilisers not taken up by crops move into rivers and grow toxic algae. In this case, dense plantings of bands of trees alongside all waterways could usefully take up the polluting surplus and turn it into a valuable product.

Humans interfere in natural cycles when they release large quantities of materials that cannot move easily into webs or networks. DDT is one of these. It simply accumulates all the way through the food chain, ending up in human fat deposits, and is not broken down to simpler substances.

Biotoxicity occurs when materials, under certain circumstances, become toxic. So, for example,

inorganic mercury found in rocks is reasonably benign until it moves into the food chain. There it is transformed into organic mercury, which accumulates in the brain and eventually destroys it. Inorganic mercury was thrown into Minamata Bay in Japan where it was absorbed into the food chain through fish. People ate the fish and 80,000 became sick or died. Today this is called Minamata disease.

Many products are marked biodegradable. Theoretically, biodegradable means a substance will be broken down into another that can move into food chains. However, very often it is broken down but then bioaccumulates because the quantity of, say, phosphates from household soap, although soluble, is simply too much for the cycle and can't be absorbed. In this way it becomes a pollutant of soils and waterways.

In permaculture we set up nutrient cycles. By using animals, compost, mulches and a selection of plants, we widen the range of materials being cycled and in some cases speed up the process.

Food chains and food webs are the structure of ecosystems

The flow of energy and the cycling of matter take place through food chains and food webs. These allow ecosystems to function. Figure 3.5 shows that from the king parrot to the earthworm is a line called a food chain. When food chains interlock they form a food web.

Together, food webs form the structure of an ecosystem. A small and weak web has very few species, few links and is vulnerable. The more complex the structure of an ecosystem, the greater its stability and strength. The more efficient the flow of energy and cycling of matter, the more likely it is to perpetuate itself. Think of a 10,000-hectare wheat field. It has a very few species and a weak structure. It can be blown down, attacked by pests, hail, drought, flood and so on. It functions inefficiently, requires huge energy inputs from the farmer, fuels and fertiliser, and can't perpetuate itself.

The key to resilience in permaculture systems is biodiversity of linked species, niches and habitats. A permaculture-designed wheat farm would have small fields protected by windbreaks of mixed species. All the necessary nutrients would be supplied by a variety of organic means, such as green manure crops and cover crops. If you don't know what these are you will find out in Chapter 6 on soils.

Succession and limiting factors

Earth has a huge range of ecosystems because of different factors acting on them. Limiting factors can be temperature, rainfall, soil, day length, altitude and distance from oceans. These tend to

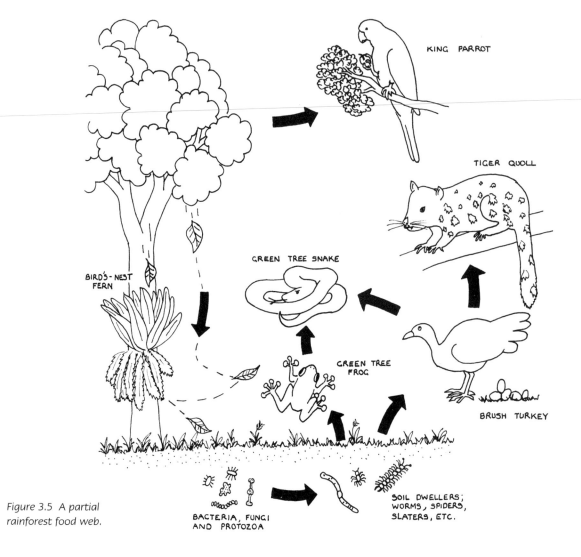

Figure 3.5 A partial rainforest food web.

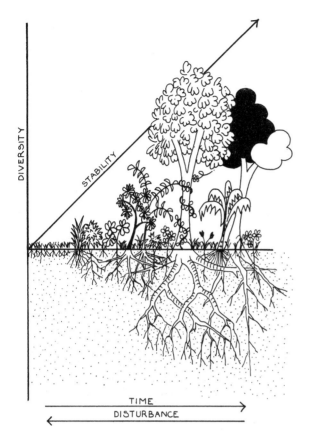

Figure 3.6 Succession.

moderate the limiting factors. Remember that every ecosystem is embedded (nested) in other eco-systems and does not exist alone. What you design will impact outside the site you are working on.

Stacking

Stacking is using space and time economically by packing in species. You can do this by:

- Planting so densely you can't see the ground and so shade out weeds. Shade-loving species are planted under tree canopies, vines are planted to climb orchard trees, and crops are inter-planted. This is called stacking in space.
- Growing different crops with each other, so that as the clover finishes the lucerne is already coming up. This is called stacking in time.

Another way to stabilise ecosystems is to use space more effectively by omitting the grass and herb succession and moving directly to planting shrubs (look again at Figure 3.6). These shrubs are known as pioneer or nurse species because they can live in degraded soil, improve soil nutrients and protect the new seedling trees. This is the main technology in rebuilding rainforests.

Your goal in designing landscapes is to move as far along the time axis and against the disturbance axis as possible.

Ecotones

The edge where two or more ecosystems meet, known as an ecotone, is extraordinarily rich and productive. Take the estuary where the ocean meets the land, or the edge where rivers meet land, or where the road meets the bush. At each of these there are many more factors of heat, warmth, water, temperature and species than in the middle of the ecosystem itself.

Ecotones impact on harvesting and design. Pest management, weed control and our relationships with wildlife are also affected. In permaculture we try to increase edge effects to create more microclimates, by providing wavy edges to garden beds, aquaculture systems and grey-water delivery techniques.

reduce, inhibit or slow down growth. Climate is the main determinant of the vegetation of an eco-system and soil is generally second in importance.

Figure 3.6 shows species diversity on the vertical axis and time on the horizontal axis. After a natural or man-made disturbance the ground will be bare. In time this will be colonised by grasses, and these will be succeeded by herbs, shrubs, small trees and bigger trees. Finally, there will be forest. This process is called succession. Succession happens because each type of plant changes the nutrient levels in the soil preparing it for the next type of plant.

Any disturbance drives succession backwards. On the other hand, as more and more species are added ecosystem stability is strengthened. For example, pioneer planting will provide windbreak protection for later wind-sensitive plants.

In permaculture you start successional planting with local species of plants and animals that do well in your region. Then, to increase the number and variety of species, you design features that

Guilds and relative placement

'Guild' is the permaculture word for co-operative groups of plants that support each other and thrive when grown together. Usually, they have evolved in the same place and under the same conditions. For example, beans, corn and pumpkins support each other. They occur naturally and also in wild systems. Acacias and eucalypts grow well together; legume and cabbage families help each other thrive. In one Aboriginal language this concept is described by the word *waru*. The bird, the worm, the tree and the spider all form a *waru*.

Relative placement occurs when you place elements relative to each other and to other parts of your overall design. For example, if you plant your vegetable garden close to your kitchen and where you walk through it every day, you will look after it better and harvest it more often. You can have your compost bin there and also send recycled water to it. When all parts save time, work and resources, the relative placement is effective. If you place your vegetable garden in one corner and your compost far off in another, the relative placement is poor and you will work harder.

The human role

We know we are consumers, but is that all we are? With our capacity to reflect, it seems as if our role could be as custodians of land and resources. Custodians 'have care of', or are keepers. This is an old familiar idea for people who live close to the earth. By taking on a new role of guardianship, perhaps we could successfully husband our Earth through this crisis.

Permaculture is not the only design system that takes into account our new understanding of the uniqueness and fragility of Earth's present biosphere and our role within it, but it is the most comprehensive.

 Try these:

The following exercises will help the theory you have just read come alive.

1. Sit in your garden and look for a food chain or part of one. If you find it difficult, watch a bird and see what it eats, or look for an eaten leaf on a plant and see if you can find the organism eating it.

2. Look in your compost bin and describe some of the inhabitants there. Draw them or just write a list of them. Are they the same in winter and summer?

3. Do seasonal counts of all the species in your garden. Start now and do it again every three months. As the numbers fluctuate you will have an indication of the stability of your ecosystem. You can draw up your page like this:

Season	Number of Animal Species	Number of Plant Species
Spring	6	30
Summer	22	31
Autumn	25	43
Winter	8	19

4. Go back to the principles of permaculture and decide which have an immediate impact on reducing your ecological footprint and how you will apply it.

5. For the next month record every single thing you bring into your garden from outside your boundaries, that is, seed, water, fertiliser, plants, mulch and so on. This is a measure of how open, or closed, your system is.

6. Do you know your ecological footprint? If not, find out now, and then ask others what theirs is.

WOW! LET ME IN

CHAPTER 4

The wonder of water

Water is life. Its management is one of the major themes of permaculture and water is likely to be the biggest issue of the twenty-first century, whether in the form of too much or too little.

Clean drinking water is no longer freely available to most of us on Earth. Rivers and lakes are usually undrinkable anywhere near cities, and treating water with chlorine and ammonia to kill pathogens means people must pay for safe drinking water and immediate health while the long-term effects of water disinfectants are unknown. Rivers, lakes, wetlands, aquifers and swamps are depleted for irrigation, housing, mining, industrial development, or are used as dumps.

Fresh water is the world's most critical resource. Some organisms can live without oxygen while none that we know of can live without water. And yet we take water so much for granted that it is difficult to realise how critical it is for our continued existence on Earth.

The amount of water in the world is finite; it cycles constantly and changes form from liquid to solid to gas. It moves from salty oceans to fresh rainwater, to ice, to rivers and soils, and back to salt water in a way that is unique in our solar system.

Of the world's total water supply, however, only 3 per cent is fresh water, and of this only 0.03 per cent is available to us at all. The rest is held in the ice caps, clouds, vegetation, aquifers and soils. The party is over. We are moving into an age of water decline—not drought, because that implies that rainfall will return to its earlier average. Instead, there is a consistent gradual decline in average

annual rainfall. However, whether your water income increases or decreases during your lifetime depends on global warming. For example, climate change, or global warming, will account for a 20 per cent increase in water scarcity in eastern Australia, and severe floods and storms along tropical eastern coasts of continents.

Because of water's essential nature, permaculture designs must obey the precautionary principle for water security and design sites as if there will be immediate and ongoing decline in water supplies and quality and that the cost of water will rise very steeply. Permaculture has two imperatives for using water:

- Live within your local water budget—don't use more than your rainfall.
- Reuse and recycle water.

 ## Our ethical task is to:

- respect all water and its origins
- accept responsibility for using water sparingly and maintaining water purity, thereby increasing ecosystem storage for the future of all Earth's community of life.

 ## Our design aims for water use are to:

- carry out a water audit
- design strategies to reduce water use and

also to use it as many times as possible before it passes out of the system
- ensure we have two or more sources of drinking and cooking water
- start implementing a water-harvesting scheme to hold water on land in soil, tanks and dams
- design water systems that rehabilitate degraded land
- tackle water problems as close as possible to their origin
- slow down water flow
- recycle and cleanse water
- create vegetated landscapes that are resistant to droughts and floods
- switch from irrigated annual crops to perennial tree crops
- harvest more clean water for future generations.

 If we don't have design aims for water use:
- we have disrespectful attitudes to water, leading to profligate overuse and pollution of it
- chemical pollutants are washed out of the sky as acid rain and pollute rivers, lakes and soils, and chemicals such as mercury, ammonia and acids enter rivers and oceans from industrial areas
- excessive use of fertilisers, particularly agricultural phosphates from agriculture, leads to ground water contamination, or algal blooms in rivers
- bacterial and viral water pollution occurs when urban water is not sufficiently aerated or lacks enough sunlight or the system is simply overloaded. Bacterial and viral toxicity can be fast and visible—typhoid and cholera—while chemical contaminants can show up as cancers or heart disease as much as 20 years later
- whole systems—animals, plants and soils—can be contaminated

- mining uses unconscionable quantities of water and pollutes rivers and ground water.

What water does for us

The environment is a legitimate user of water. When we try to dam it or drain it we are depriving the natural environment and eventually this will affect all of us. Nature uses water in wetlands, rivers, swamps, soils and ground water to regulate vegetation and climate and drought-proof the land. Modern enterprises reverse this.

There is no other liquid whose functions work for us in as many ways and forms as water does. Table 4.1 shows just a few.

The supply and release of water in many forms is inextricably linked. Damage to one part of the cycle affects others. For example, forest clearing results in rising water, which results in soil salinity, which results in desertification. Melting ice caps and glaciers raise saltwater levels affecting coasts and fish-breeding grounds.

TABLE 4.1: THE FUNCTIONS OF WATER

Form	Function
Gas	carries heat around the world humidifies airstreams
Liquid	is the basis of nutrition and absorption of nutrients for plants and animals is present at the conception of every life form is the universal solvent provides a home for many animals gives renewable energy is a fine transport system is important for health
Solid	it is a vast reservoir of fresh water it carves out valleys and carries huge loads it prunes trees it melts and allows whole nations to have running water for many months of the year as ice, it is a preservative assists climate stability

Carrying out a water audit

To ensure water security you need to have enough water from an assured source all year round for drinking, cooking and all forms of washing. The garden should be able to be watered from grey water. In permaculture every major theme, such as water, energy or food, requires you to design two or more separate sources in case one gets damaged, polluted or even stolen. So, if you are using tank water, then you should have two smaller tanks instead of one big one. You could also have town water and a tank, or a protected dam and a tank, and so on. Remember that if you control what goes into it then even dam water can be filtered and boiled. Generally, you can think about water as being first-class for drinking and cooking and second-class for showers, toilets, gardens and washing.

Water sources and risks

- Town supply can be vulnerable to loss, depletion, toxins, rationing and uncontrollable price increases.
- The quality of tank or rain water may be uncertain in cities yet you can substitute it for town supply in gardens and washing machines, to clean the car, wash the dog and flush the toilet.
- Spring water may dry up.
- River water is not reliable or safe.
- Bore water belongs to the earth and carries out special functions; it can also turn salty or dry up. The use of bore water has resulted in arsenic problems in South-East Asia and many other areas, and salinity problems in east coast India. In the future, most governments will meter bores and charge for water. UNICEF has reversed its single focus from bores as the only source of water to bores as supplementary water supplies in an effort to relieve bore water problems in developing countries.
- Snowmelt canals may be intercepted and the water polluted uphill from where it is used.

Water tanks are now mandatory in New South Wales, Australia, for all new houses. When tank water is plumbed to washing machines and toilets the water authority gives a substantial rebate on the cost of the tanks. Figure 4.1 illustrates how water for city or suburban homes enters clean and leaves dirty.

How to calculate your water security needs

To have water security you need to know:

- how much you get from rainwater
- how much you use
- how much you must store for the longest drought.

Most of us don't know how much water we use. We don't know how much falls on our land each year—our annual rainfall—and its distribution. Are we using more than falls? If so, will we eventually run out? We also don't know where and how water is disposed of once we've used it. Answering these questions requires you to perform what's called a water audit. You measure how much water comes in and how much goes out of your site. This will tell you whether you are living beyond your water means or whether you are using water sustainably.

If you rely on town supply you can expect the cost of water to increase significantly in the next few years and, if it hasn't already done so, for rationing to start. By ensuring your own water security now you are protecting yourself from price rises and the uncertainty of quantity and quality of supply, and you are conforming to the permaculture principle that encourages self-reliance.

To find how much water you use per day you can look up your last water bill, or you can do as Rob did (see Figure 4.2) and calculate it bucket by bucket, or as Leanne did (see Table 4.3) and study your water meter. Rob and Leanne, a permaculturist from Leura, both saw where water use was the greatest in their households. Figure 4.2 shows Rob's tally for a week and he also compares it with the average use for Sydney. With this he knew where to start to reduce his water use and which changes would make the greatest saving.

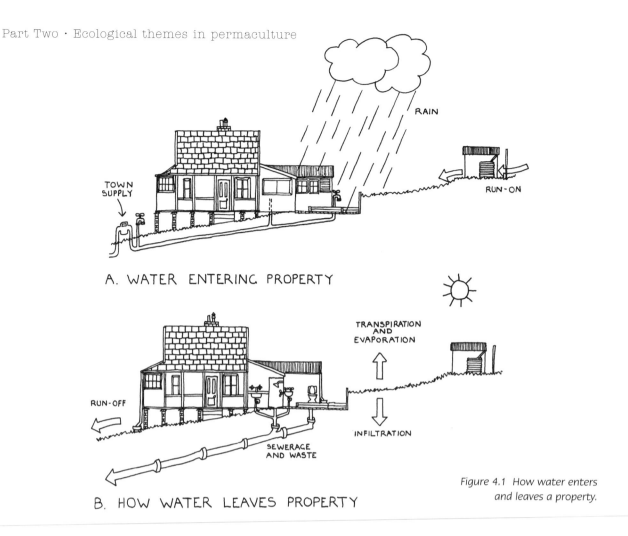

A. WATER ENTERING PROPERTY

B. HOW WATER LEAVES PROPERTY

Figure 4.1 How water enters and leaves a property.

You also need to know how much water you can harvest. And to find this out, you need to do some simple calculations. You need to know your average annual rainfall, your roof area and the longest period without rain. Table 4.2 shows what I calculated for my household of three people. I have equipped myself against a six-month drought and still live well. In the severe Sydney drought of 2003 all my neighbours had to buy in water, which was expensive, while I still had one tank left when the rains came.

Figure 4.2 shows how to calculate your storage needs. Now do your own calculations for the storage you require to ensure your water security. Everyone should know how much water they use daily and where it comes from and where it goes.

How much should you use?

The World Health Organization (WHO) says we need 2 litres a day for personal consumption. And to wash clothes, dishes and ourselves, we need 45

litres a day. Californians use 1000 litres a day, and until recently West Australians were using about this amount. They got it from bore water which eventually ran out and now their situation is extremely serious.

So, what can you do if you are using more water than you harvest? You can:

- Increase your roof catchment by increasing the roof area or catch surplus off the roof and direct it to dams or ponds.
- Decrease your water use by using less, or you can clean and reuse water known as 'grey water'. Both Israel and Singapore have refined processes to clean grey water to a high standard of purity and reuse it. Singapore water is said to pass through seven people.

Grey water and tank overflow

These are both under-utilised resources. For example, my calculations show that I have 120

WATER AUDIT

	DAY 1 MON. 21/9	DAY 2 TUE 22/9	DAY 3 WED 23/9	DAY 4 THUR. 24/9	DAY 5 FRI 25/9	DAY 6 SAT 26/9	DAY 7 SUN 27/9	AVERAGE DAILY TOTALS	SYDNEY AVERAGE
DISH WASH 8 lt. / WASH	II (16 lt)	II (16 lt)	III (24 lt)	—	II (16 lt)	III (24 lt)	I (8 lt)	15 lt	UP TO 18 lt./WASH
SHOWER 10 lt./ MIN.	I (AV. 7 min) (70 lt)	I (70 lt)	I (70 lt)	I (70 lt)	—	I (70 lt)	I (70 lt)	60 lt	UP TO 250 lt./ SHOWER
TOILET 11 lt. / FLUSH	IIII (44 lt)	III (33 lt)	III (33 lt)	II (22 lt)	III (33 lt)	HHT (55 lt)	IIII (44 lt)	38 lt	UP TO 13 lt./ FLUSH
HAND WASH 2 lt./ WASH	HHT (10 lt)	HHT (10 lt)	IIII (8 lt)	III (6 lt)	IIII (8 lt)	HHT (10 lt)	HHT (10 lt)	91 lt	UP TO 5 lt./WASH
TEETH CLEAN 1 lt. / CLEAN	II (2 lt)	II (2 lt)	II (2 lt)	II (2 lt)	II (2 lt)	III (3 lt)	III (3 lt)	2 lt	UP TO 5 lt./ CLEAN
COOKING DRINKING X 1 lt.	IIII (4 lt)	III (3 lt)	IIII (4 lt)	II (2 lt)	III (3 lt)	HHT (5 lt)	HHT (5 lt)	4 lt	8 lt./ DAY
WASHING MACHINE 150 lt./LOAD	—	—	—	—	—	II (300 lt)	—	42 lt	UP TO 265 lt./ WASH
GARDEN UP TO 25 lt./MIN.	—	—	HOSE FOR 10 MINS (200 lt)	—	—	SPRINKLER FOR 30 MIN (600 lt)	—	114 lt	UP TO 1,500 lt./ HOUR
MISC.			(5 lt)			(10 lt)		3 lt	
TOTALS	146 lt.	134 lt.	346 lt.	102 lt.	62 lt.	1077 lt.	145 lt.	287 lt./DAY	

Figure 4.2 Water audit for Rob's household.

litres of grey water per day (840 litres of grey water a week) passing through my house. This grey water comes from the washing machine, bathroom and kitchen. The toilet water, called black water, is excluded. So I have about 500 litres a week available for the garden, glasshouse or other uses. Grey water from washing machines can be plumbed to the garden though soaker hoses under mulch. While 500 litres is enough water for a well-managed garden in a drought, it is not easy to

TABLE 4.2: THE AUTHOR'S WATER AUDIT

Data needed	My answers
1. The annual average rainfall	1400 mL/year
2. The longest dry period	six-month drought (340 days), not normal, probably climate change
3. Roof catchment	390 m² of roof
4. Total potential roof catchment/year	390 x 1400 mL = 546,000 L
5. Daily water use	120 L/day
6. Total use/year	120 x 365 = 43,800 L
7. Water needed for six months	120 L/day x 183 days = 21,960
8. Water storage required for water security of six months (longest dry period) + 20% contingency	22,000 + 4000 L (say, 1 x 10,000 L and 1 x 15,000 L tanks, or 2 x 15,000 L tanks, or any approximate combination

CALCULATING YOUR WATER STORAGE NEEDS

- FROM YOUR WATER AUDIT, CALCULATE THE FOLLOWING,

  HOUSEHOLD CONSUMPTION PER DAY   __287 lt.__

  HOUSEHOLD CONSUMPTION PER WEEK   __2012 lt.__

  HOUSEHOLD CONSUMPTION PER YEAR   __104,624 lt.__

- NEXT, FIND OUT THE AVERAGE ANNUAL RAINFALL FOR YOUR AREA

  __1300mm__

- CALCULATE THE SURFACE AREA OF YOUR ROOF CAPABLE OF CATCHING RAINFALL FOR STORAGE

  __91·5m²__

- IF YOU MULTIPLY YOUR AVERAGE ANNUAL RAINFALL BY THE SURFACE AREA OF YOUR ROOF, THIS IS IN THEORY THE AMOUNT OF RAINFALL IT IS POSSIBLE FOR YOU TO CATCH AND STORE EACH YEAR

  RAINFALL X ROOF AREA __118,950 lt./YEAR__

- FIND OUT THE AVERAGE LONGEST PERIOD OF TIME BETWEEN GOOD FALLS OF RAIN

  __2 MONTHS__

- IF YOU NOW WORK OUT HOW MUCH WATER YOU WOULD USE DURING THIS DRY PERIOD, YOU CAN CALCULATE HOW MUCH WATER YOU NEED TO STORE

  DRY PERIOD X WATER CONSUMPTION __16,096 lt.__

  (NOTE. THIS IS A MINIMUM AMOUNT AS IT IS ALWAYS BEST TO HAVE A SURPLUS FOR UNFORSEEN EMERGENCIES AND SEASONAL VARIATIONS)

- ESTIMATE THE SIZE OF TANKS OR STORAGE FACILITIES NEEDED TO HOLD THIS AMOUNT   __2x 10,000 lt. tanks or(2x 4,000 Gal. tanks)__

Figure 4.3 Calculating your water storage needs. This example has been filled in for Rob's place.

TABLE 4.3: LEANNE'S FAMILY'S WATER CONSUMPTION

Household consumption per day	600 L	Minimum annual rainfall	856 mm
Household consumption per week	4200 L	Maximum annual rainfall	1941 mm
Household consumption per year	219,000 L	Average annual rainfall	1400 mm

Surface area of roof: House 166 m²; Garage 24 m²; Chicken shed 7.5 m²

Total annual rainwater collecting capacity = rainfall x roof area

	Minimum	Maximum	Average
House	142,096 L/yr	322,206 L/yr	232,400 L/yr
Garage	20,544 L/yr	46,584 L/yr	33,600 L/yr
Chicken shed	6,420 L/yr	14,557 L/yr	10,500 L/yr

Storage for three months collected from house roof = consumption x time 54,700 L

Size of tanks or storage facilities: 3 x 20,000 L tanks; 2 x 30,000 L tanks

recycle 500 litres of grey water when it's raining. Combined with the average rainfall it is simply too much, especially if you are aiming for zero run-off or clean water run-off. This is a cogent reason to reduce your water consumption. Huge amounts of grey water from villages, towns and cities pollute rivers and oceans because the environment cannot effectively dilute and cleanse the surplus water. I endeavour to manage all my own waste, including grey water, by copying nature.

Rob redirected surplus water to his garden and increased biomass (or plant) infiltration and water cleansing by creating gutters in his driveway and contouring winding paths. He stored water in tanks to use in the dry periods. You will save energy if you place your rainwater tanks high off the ground so they can gravity-feed. If you use a pump, then have a high header tank and only pump up to it once a day.

A ten-year water audit by an Australian family

Table 4.3 shows the water audit calculations for Leanne's family of two adults and two teenagers. Based on these figures they implemented changes aimed at reducing their water consumption. Leanne said, 'The most important factor relating to all the decisions we have made is that all the water leaving our land goes directly into the Sydney catchment via Gordon Falls. We have therefore endeavoured to keep sediment and nutrient run-off to an absolute minimum.'

Leanne identified the following factors contributing towards high water use:
- no measures to control evaporation from garden beds
- high usage of a top-loading washing machine, using 300 litres per day
- no water-saving devices installed
- no regulations for length of showers.

Leanne kept records for ten years of her family's water use and its cost. Since she implemented water-saving methods at home her usage has gone down 30 per cent while her costs have stayed virtually the same because the price of water has increased substantially during that decade. Water consumption was very high initially, at a time when the cost of water was relatively cheap.

The family's water consumption began to fall after the initiation of the following strategies:
- Over ten years large garden beds have been established and continually mulched to retain moisture. To date Leanne has distributed 90 cubic metres of chippings, 57 bales of lucerne, continual recycling of compost and the chipped mulch from their own prunings.
- Plant selection has been aimed at choosing the right plant for the conditions.
- A front-loading washing machine was installed bringing usage down to 65 litres per day.
- Water-saving devices were installed: a dual flush toilet cistern (4.5 litres and 9 litres) and

reduced-flow shower rose (10 litres per minute).

- They only wash the car after a holiday at the coast (maybe once per year).
- The only plants watered are the newly planted shrubs until established and the vegetables and lemon tree.

According to Leanne, 'Around the year 2000 water consumption began to rise again! This coincides with the time that my two children decided that they didn't want to share a bath any more and would prefer to shower. This trend continued until 2004 when timed showers were enforced and with huge success. The consumption of water has continued to fall. Despite the drought, the garden is flourishing.'

Leanne has developed a 'grand plan' to ensure that water consumption will continue to be kept to a minimum.

- Use the 'if it's yellow let it mellow, if it's brown flush it down' approach to toilet flushing.
- Reduce shower times to three minutes (30 litres), and five minutes (50 litres) for hair washing.
- Ensure that the tap is not left running during teeth cleaning (reducing water use from 2 litres to 250 millilitres).
- The installation of a water tank for the house.
- The installation of gutters and a water tank for the chicken shed. The captured water supplies the vegetable garden, chickens and the duck pond.
- Investigate the feasibility of grey-water processing.
- Continue to top up the garden with mulch. (Chippings are purchased cheaply from a local tree lopper.)

Do these calculations for yourself using mine, Rob's or Leanne's calculations as a model. Would you have water security if the town supply failed or if there was a long and severe drought?

Reducing water consumption and still living well

A primary permaculture water principle is to reduce water consumption and reuse all water as many times as possible before it passes out of your system. Table 4.4 lists some water-saving strategies and you can think of others.

Cleaning and reusing domestic water

Seventy per cent of Israel's municipal grey water is treated and reused in agriculture. There are also some simple strategies for reusing domestic water and cleaning it. These rely on copying nature and lead to the second permaculture principle for water: ensure water is biologically filtered and cleansed by your system before it leaves your land.

Nature cleans water by:

- slowing it down so it can drop some of its load
- filtering it through mulches and soil
- passing it over river stones to oxygenate it
- sterilising it with sunlight
- feeding it through vegetation.

To filter and clean water before it leaves your land, use the following techniques:

- When you replumb grey water to the garden, deliver it slowly and under mulch.
- Put only into your household water what you want to eat—simple pure, zero- or low-phosphate soaps.
- Use the fall of the land to move water by gravity.
- Store water in very wet weather until you need it.
- Hot bath water can be stored, used to heat rooms and greenhouses by plumbing it to these places or switched to washing machines. By placing baths higher than washing machines or washing machines below baths, bathwater can be reused as the first clothes wash.
- Fit filters at the end of outgoing hoses and clean them regularly.
- If you store water in ponds, keep them

TABLE 4.4: DOMESTIC WATER STRATEGIES

Consumption	Saving techniques
Kitchen	One washing-up per day in a basin with the water emptied on the garden. Plumb the sink drain to the garden. Keep a basin in the kitchen sink for every rinsing and use it for the next pre-wash. Send kitchen grey water to orchards. In rural areas use rainwater only for drinking and cooking. Fit low-delivery taps.
Laundry	Only use washing machines with a full load. Wash clothes less often. Fit low-delivery taps. Plumb laundry water to garden biomass (plants).
Bathroom	Fit low-delivery taps and roses on showers. Re-plumb the handbasin to the toilet or garden. Fit push-button showers, or timers set for 3–5 minutes. Keep the plug in the handbasin and wash hands several times before sending the water to the garden. Fit half-flush toilets, or put a brick in the cistern, or bend the float arm to a lower level. Install a compost toilet.
Other	Mulch the garden. Bucket-wash cars. Fit new washers to dripping taps. Water the garden by hand. Turn the swimming pool into aquaculture and swim at beaches and rivers.

shallow so the sun can act as a steriliser and the wind can oxygenate it.

● The best way to reuse domestic water is to store it in biomass (the plants and animals in your garden). A diverse and densely planted backyard garden or a well-forested farm will store much more water in biomass than a lawn or bare field. This water can be harvested as mulch, fruit, firewood, etc. The systems that you design and the strategies you implement should increase water storage and yields for many years (see Figure 4.4).

Both soil and plants can clean water if you don't use too many chemicals. So make your shopping simpler, and your household safer, by buying and using the simplest soaps you can find. Make your own washing liquid for washing up and using in washing machines, and use vinegar in dishwashers.

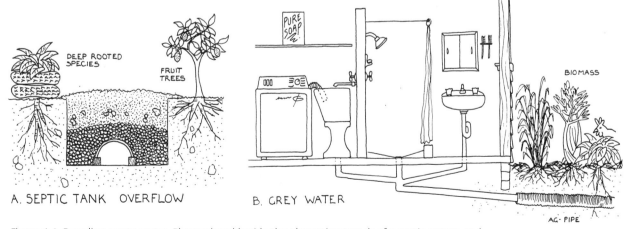

Figure 4.4 Recycling waste water. Plants placed beside the absorption trench of a septic system and greywater outlets assist in cleaning water (after B. Mollison, Introduction to Permaculture, *p. 111).*

Making a whole site water plan

You need to design a whole site water plan to achieve your objectives for water and not waste time and resources. This is the macro-design and involves placing all the water on your land so it works harmoniously. You will need to think how water moves across your land, is stored and leaves it, and what would be ideal and sustainable in the future. The present water storage and use is your water analysis, and the design you draw up is your sustainable plan. To do this you need to know more about water than just your water audit.

On your base plan show your whole site water analysis with the following features:

- How it comes in—and the risks and quality of that water.
- Where it comes in, for example, chicken-shed roof, garden shed, house, garage, and so on.
- How it is stored—44-gallon drums, tanks and their size, ponds, etc.
- Where it is used—bathrooms, laundry, kitchen (the wet areas).
- The quality and quantity of grey water—700 litres per week with only bland soap, shampoo, etc.
- Where the grey water goes—reed bed, gardens, underground, etc.

Look at my whole site water plan. I have done a plan view, that is, a bird's eye view (see Figure 4.5), so you can see the layout of the rooms, and an elevation or side view (see Figure 4.6). Draw a similar plan for yourself. What strategies and techniques do you use to reduce your water consumption? Keep records of your water consumption.

Water strategies for rural land

Water is the primary selection factor in choosing land. This means you need to know the rainfall, its distribution, run-off, streams, dams, rivers and watershed control before you buy. If available water is too little, then farming will be very discouraging. If the water is contaminated, you will want to leave.

Always plan your farm enterprises so you live within your water budget—your annual average rainfall. The following permaculture strategies will help you do this.

When you are working with water on farmland the unit of self-sufficiency and repair is the watershed. The very best water management comes from farmers working together across a contoured landscape to store and move water so as to benefit all the downhill slopes. This is also the best way of rehabilitating farmland. Working on their own, farmers risk being sabotaged by other non-participating farmers. Farmers need enough water for all and the revegetation and sharing of a watershed provides benefits that are beyond the individual farmer's ability to implement.

A farmer's two main priorities are to manage water and to store water.

The key objectives for managing water are to:

- try to let no run-off water escape from your property until all storages are full
- slow down water flow
- use water as many times as possible
- tackle excess water problems as close as possible to where they originate—the top of your path, or the top of your watershed
- clean water by passing it through biological filters or traps.

The key objectives for storing water are to:

- maximise water stored in soils, because it is the most efficient storage and requires least energy use
- maximise water in biomass—the plants and animals—because this form of storage is most efficiently harvested
- use Yeoman's Keyline (see page 41) to hold surface water held in dams and ponds.

These techniques can be used for small or large areas of land. On large rural areas, machinery is used, and in gardens, hand tools.

Increasing water in soils

Most soils you will work with hold a small percentage of the water they held originally under natural vegetation. Your priority is to increase water

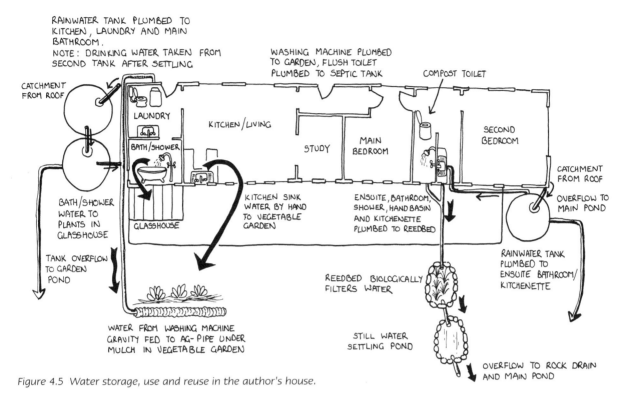

RAINWATER TANK PLUMBED TO
KITCHEN, LAUNDRY AND MAIN
BATHROOM.
NOTE: DRINKING WATER TAKEN FROM
SECOND TANK AFTER SETTLING

WASHING MACHINE PLUMBED
TO GARDEN, FLUSH TOILET
PLUMBED TO SEPTIC TANK

COMPOST TOILET

CATCHMENT
FROM ROOF

LAUNDRY

KITCHEN/LIVING

STUDY

MAIN
BEDROOM

SECOND
BEDROOM

BATH/SHOWER

CATCHMENT
FROM ROOF

OVERFLOW TO
MAIN POND

GLASSHOUSE

BATH/SHOWER
WATER TO
PLANTS IN
GLASSHOUSE

KITCHEN SINK
WATER BY HAND
TO VEGETABLE
GARDEN

ENSUITE, BATHROOM,
SHOWER, HAND BASIN
AND KITCHENETTE
PLUMBED TO REEDBED

RAINWATER TANK
PLUMBED TO
ENSUITE BATHROOM/
KITCHENETTE

TANK OVERFLOW
TO GARDEN
POND

REEDBED BIOLOGICALLY
FILTERS WATER

WATER FROM WASHING MACHINE
GRAVITY FED TO AG-PIPE UNDER
MULCH IN VEGETABLE GARDEN

STILL WATER
SETTLING POND

OVERFLOW TO ROCK DRAIN
AND MAIN POND

Figure 4.5 Water storage, use and reuse in the author's house.

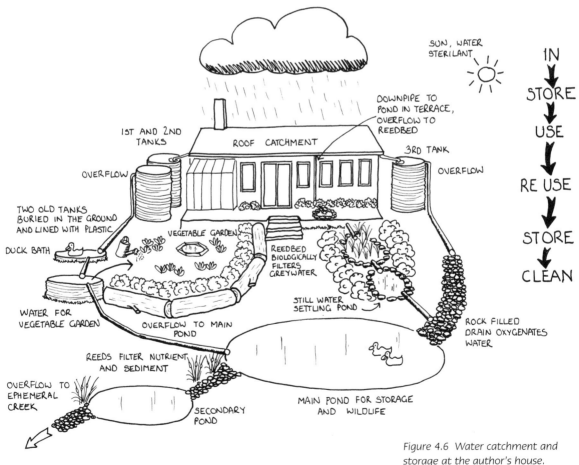

SUN, WATER
STERILANT

IN
STORE
USE
RE USE
STORE
CLEAN

1ST AND 2ND
TANKS

DOWNPIPE TO
POND IN TERRACE,
OVERFLOW TO
REEDBED

ROOF CATCHMENT

3RD TANK

OVERFLOW

OVERFLOW

TWO OLD TANKS
BURIED IN THE GROUND
AND LINED WITH PLASTIC

VEGETABLE GARDEN

REEDBED
BIOLOGICALLY
FILTERS
GREYWATER

DUCK BATH

WATER FOR
VEGETABLE GARDEN

OVERFLOW TO MAIN
POND

STILL WATER
SETTLING POND

ROCK FILLED
DRAIN OXYGENATES
WATER

REEDS FILTER NUTRIENT
AND SEDIMENT

OVERFLOW TO
EPHEMERAL
CREEK

SECONDARY
POND

MAIN POND FOR STORAGE
AND WILDLIFE

*Figure 4.6 Water catchment and
storage at the author's house.*

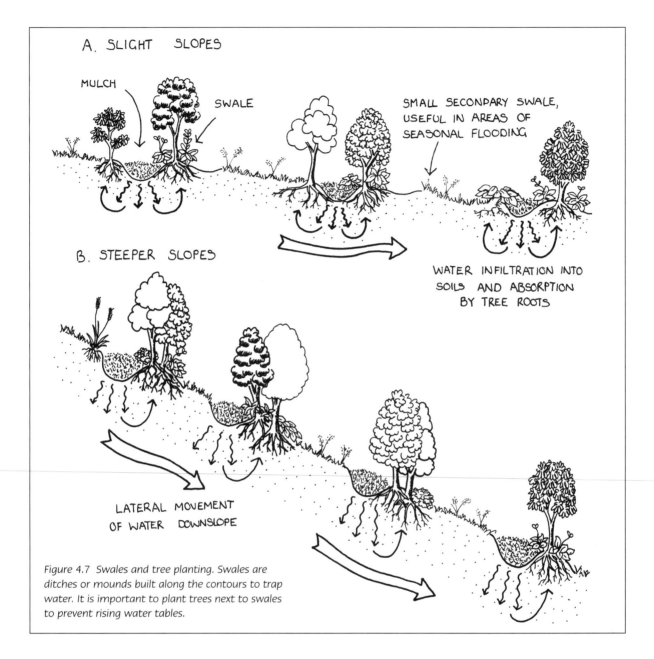

A. SLIGHT SLOPES

MULCH

SWALE

SMALL SECONDARY SWALE, USEFUL IN AREAS OF SEASONAL FLOODING

WATER INFILTRATION INTO SOILS AND ABSORPTION BY TREE ROOTS

B. STEEPER SLOPES

LATERAL MOVEMENT OF WATER DOWNSLOPE

Figure 4.7 Swales and tree planting. Swales are ditches or mounds built along the contours to trap water. It is important to plant trees next to swales to prevent rising water tables.

in soils and their water-holding capacity. This is also the first step in the rehabilitation of soils.

Begin by trapping water as high as possible on the land by ripping the subsoil deeply without turning over the topsoil (see Figure 4.7). Then plant into the rip lines. Trees shed about 25 per cent of their root system each year and this, together with soil micro-organisms, becomes organic matter, which in turn holds large amounts of water in the soil.

Next, make swales. These are ditches that slow water as it flows downhill, giving it time to be absorbed. They are constructed along the contours

of the land with an A-frame or by survey, and any overflow water from one is caught by the next swale below it (see Figure 4.8). However, in principle, if swales overflow then there are not enough of them. The building of swales is site specific and these general rules apply:

- The steeper the slope the closer the swales (almost terraces or steps).
- The less cohesive the soil structure the further apart the swales.

Swales recharge the soil's ground water, resulting in 85 per cent less run-off than from bare

YOU NEED THREE FLAT PIECES OF TIMBER OF EQUAL LENGTH, A HAMMER, NAILS, STRING AND PLUMB BOB.

MAKE AN A-FRAME. MAKE SURE THE ANGLES BETWEEN THE LEGS AND THE CROSS-PIECE ARE THE SAME AND THE LEGS ARE OF EQUAL LENGTH. TIE THE PLUMB BOB FROM THE TOP TO CROSS THE A.

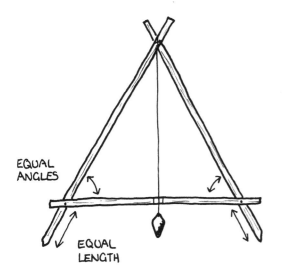

EQUAL ANGLES

EQUAL LENGTH

PLACE BOTH ITS FEET ALONG A CONTOUR UNTIL THE PLUMB BOB CROSSES THE MIDDLE. MARK THE PLACE OF ITS FEET WITH SMALL STAKES. NOW SWING THE WHOLE FRAME ACROSS TO A NEW POSITION AND MARK IT WHEN THE PLUMB BOB HAS CENTRED. CONTINUE UNTIL YOU REACH YOUR BOUNDARY. LEAVE THE MARKING STAKES IN PLACE UNTIL YOU DIG YOUR SWALE.

ON A FARM, MARK THE LINE WITH LIME OR FLOUR AND THEN DIG YOUR SWALE WITH MACHINERY. OR YOU CAN PAY A SURVEYOR TO FIND YOUR CONTOURS.

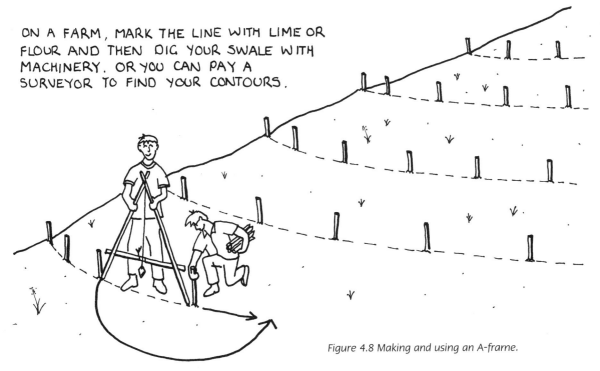

Figure 4.8 Making and using an A-frame.

land and a 75 per cent increase in the soil's ability to retain water. After the first rains dams and rivers will not fill or flow until the soil water is recharged. The surplus water moving to lakes, dams and rivers will then be clean because the water will have effectively dropped or filtered its load of sediment which it held. In many cases, after several years of good rain, the water will eventually break through lower down the slope, as a spring.

Storing water in soils and accompanying it with tree planting is the first step in drought-proofing your land.

Preventing evaporation by using mulches

To prevent the water held in the soil from evaporating the soil needs a cover. In fact you should feel uneasy when you see bare land—it's like skin with a layer missing. The most effective way to protect soil is to use mulches.

Mulch is a layer over the soil which protects it from the damaging effects of wind, sun and water. Mulches have several special functions and multiple benefits:

- They reduce soil evaporation and therefore inhibit soil salination and general water loss.
- They increase water infiltration by absorbing water on the surface and holding it until it has time to be absorbed.
- Mulches reduce erosion from gravity, wind and water.
- They regulate soil temperatures by reducing extremes of summer heat and winter cold.
- They suppress weeds, which also rob the soil of water.
- They can raise the light in a dark area or reduce light intensity and increase soil warmth.
- Mulches supply nutrients and organic matter to soils.
- They are one way of using up your garden surplus.

Figure 4.9 gives examples of some types of mulches and how they can be used.

Storing water in biomass

All living things are about 80–85 per cent moisture. When you increase water stored in the soil, you

INORGANIC/ SYNTHETIC	ORGANIC	
	LIVING	DECEASED
USED IN ARID LANDSCAPES AND URBAN AREAS (SHORT TERM)	USED IN ZONE II, BROADSCALE AREAS, AND ZONE I IN HUMID TROPICS	USED IN ZONE I AS SHEET MULCH AND IN ZONE II AS SPOT MULCH
CARPETS BLACK PLASTC CORRUGATED IRON WEED MAT STONES GRAVEL RIVER SAND PATHS	COVER CROPS LEGUMES CLOVERS PUMPKINS POTATOES SWEET POTATOES CLOSELY PLACED HERBS SELF-MULCHING SPECIES NATIVE GROUNDCOVERS	MANURES STRAW LEAVES SEAWEED LAWN CLIPPINGS NEWSPAPERS WOODCHIP FOOD SCRAPS ANIMALS DENIM WOOLLEN CARPETS SAWDUST PINENEEDLES COMPOST

Figure 4.9 Types of mulches.

increase the quantity and variety of living organisms per hectare. You can also stock more densely, thereby increasing the diversity of productive plants in your landscape. This will increase productivity for many years. Water stored in biomass can be harvested as mulches, fruit, vegetables, grains, oils, dyes, juices, eggs, meat and fibres.

Surface harvesting and storage of water on farms

In tropical and cool temperate climates, having 15 per cent of water stored as surface water will reduce fire risk, provide a buffer against climate change and modify climate extremes. On seaward slopes, where it may not rain much yet the air is humid, fog fences—like fishermen's nets—hung over sheets of collecting plastic will harvest the water at night as it condenses and runs into containers. Sometimes roofs act like water condensers when the air is moist and the cool water will drip into your tanks.

Usually farmers place small, exposed dams at the bottom of the slopes and these dry up when farmers need them the most—in dry seasons or drought. It is the water-harvesting design strategy called Yeoman's Keyline Water Harvesting which is the most reliable water strategy and assists with whole farm planning throughout the watershed. This strategy is safe, environmentally friendly, relatively inexpensive and provides farmers with water security.

Yeoman's Keyline Water Harvesting

The Yeoman's design begins uphill with a study of the land's contours. The tops of slopes are the driest and most difficult to rehabilitate, and by starting high up on the land the flow-on effects downhill are greater. The first and most important dams are built high up on the land, and are kept clean and protected with close plantings of trees and shrubs around them. The water stored in these high dams is distributed by gravity to farm enterprises below them. The dams are linked into a network by a series of contour banks which, like swales, run mainly along contours. The contour banks have

some fall and carry water along ditches from one dam to another surface catchment dam. These dams are also as high up on the land as possible. At every stage of the building of swales and contour banks, trees are planted. Water and trees go together like salt and pepper.

When a whole watershed has been keylined, the landscape becomes a series of dams linked by contour banks. No dam is very far from another. You will find details of this strategy in the book *Water for Every Farm: The keyline system*, by Allan Yeomans.

Low dams are used to hold grey water. They act as a productive aquaculture water system which, with its associated water plants and aquatic animals, cleans water and filters some toxins and excess nutrients before the water overflows and rejoins rivers or lakes.

In Australia dams are not permitted on rivers or creeks because they will inhibit environmental flows. Therefore, to store water you can run a channel or swale from the riverbank to a dam built parallel to the river course. During floods water will flow into it and be harvested in the dam. You are also prohibited from digging dams with a surface area of 1 hectare or more. The over-construction of many of these dams has prevented the environmental flows that are required to keep rivers flushed out and clean. Yeoman's strategy enables you to design many small dams across your land and you will have water security and total control over the use of your water. This design will serve you better because you can place dams close to the enterprises where you want to use water. An added benefit is that if one dam is polluted or goes dry you will still have multiple sources available (see Figure 4.10).

Biological water cleansing

Plants on the edges of lakes, swamps and rivers act as natural filters of dissolved chemicals and physical matter such as clay particles. Try to observe a natural ecosystem where this is happening and copy it. By copying these natural aquatic ecosystems, a biological filtration system can be

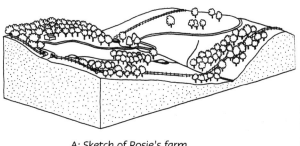

A: Sketch of Rosie's farm
before water harvesting

B: Plan vew after
implementation

Figure 4.10 Surface capture of water on Rosie's farm. Water is trapped on the high parts of the property and is gravity fed to the lower dams. Plants which can filter silt and chemical residues are planted around the dams and along the boundaries to prevent contaminated water entering the property (after B. Mollison, Permaculture: A Designer's Manual, p. 18).

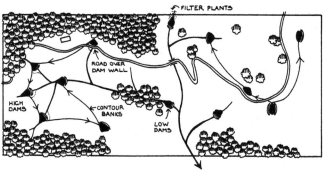

constructed to treat grey water. Nutrient-rich grey water enters the system at one end and travels through a series of ponds which gradually filter and remove solid material and dissolved nutrients. The plants growing in the ponds can also be harvested for mulch.

The topic of grey water and its safety, effects on soils and risks to human health are completely covered in Echo Development Notes of July 2005 (see References). The main conclusion, after covering exceptions, is that grey water should always go through compost or mulches if it is to be safe. To use more comprehensive systems you can build a biological water filter which mimics a wetland.

A biological water filter system is based on the elements required to clean water naturally. The size of the system depends on the amount of water passing through it and the season. Some macrophytes (wetland reeds) die down in winter and the system also works more slowly. For a household, the system should be about 5 metres long and 2 metres wide. Figure 4.11 shows one possible design. The water leaving a bathroom is dropped 2 metres deep into a pond and slowly filters upwards before it moves into the pools. It moves through:

- gravel and sand to filter non-soluble materials like lint, seeds and clays

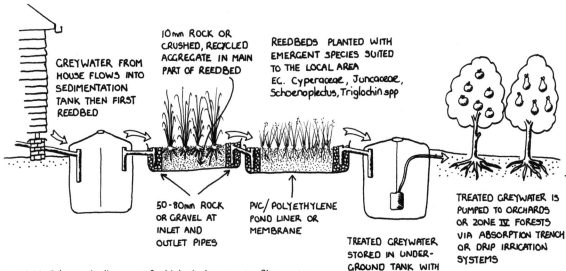

Figure 4.11 Schematic diagram of a biological greywater filter system using reedbeds (refer to detailed diagrams on opposite page).

REEDBED DETAIL

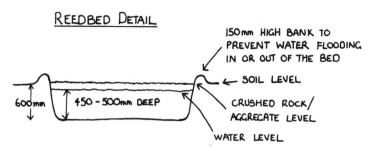

150mm HIGH BANK TO
PREVENT WATER FLOODING
IN OR OUT OF THE BED

← SOIL LEVEL

CRUSHED ROCK/
AGGREGATE LEVEL

WATER LEVEL

600mm   450 - 500mm DEEP

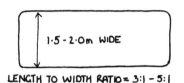

1.5 - 2.0m WIDE

LENGTH TO WIDTH RATIO = 3:1 - 5:1

AREA OF REEDBED NEEDED PER PERSON
BASED ON GREYWATER SOURCE

- WATER FROM LAUNDRY   1.5m² / PERSON
- BATHROOM / SHOWER   2.0m² / PERSON
- ALL HOUSEHOLD GREYWATER   4.0m² / PERSON
- COMBINED WASTEWATER   6.0m² / PERSON
  (INCLUDING TOILET)

ABSORPTION TRENCH DETAIL

ABSORPTION TRENCHES MAY HAVE SINGLE
OR DOUBLE DISTRIBUTION PIPES

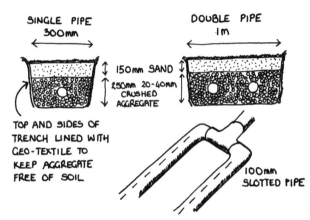

SINGLE PIPE
300mm

DOUBLE PIPE
1m

150mm SAND

250mm 20-40mm
CRUSHED
AGGREGATE

TOP AND SIDES OF
TRENCH LINED WITH
GEO-TEXTILE TO
KEEP AGGREGATE
FREE OF SOIL

100mm
SLOTTED PIPE

DRIP IRRIGATION DETAIL

TREATED GREYWATER CAN ALSO BE
PUMPED THROUGH DRIP-IRRIGATION
SYSTEMS

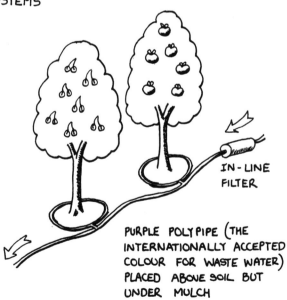

IN-LINE
FILTER

PURPLE POLYPIPE (THE
INTERNATIONALLY ACCEPTED
COLOUR FOR WASTE WATER)
PLACED ABOVE SOIL BUT
UNDER MULCH

FLOW-FORMS

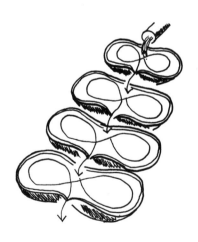

A SERIES OF FLOW-FORMS MAY
BE INCLUDED IN THE GREYWATER
FILTERING SYSTEM AFTER WATER
HAS PASSED THROUGH REEDBEDS.
FLOW-FORMS OXYGENATE WATER

- bulrushes to absorb undesirable chemicals through their root systems
- lotus, waterlilies and water hyacinth to absorb specific pathogens or chemicals and to cool the water
- unshaded water pools so the sun can act as a steriliser and the wind can oxygenate the water
- water movement from one pool to another as the water is cleaned.

The importance of rivers, ponds, lakes wetlands, bays, aquifers and soil moisture

Rivers are spirit paths. Songs, birds, water and trees travel along river valleys. If you have these on your land then you are lucky. If you haven't, then try to find even a small place for an artificial water body. A recent study of water across continents and its effect on climate showed that 'wetness' of the land is a critical factor in maintaining climate stability (*The Science Show*, Radio National, CSIRO report, February 2005). The study found that it is the wetness of the soil and the area of water bodies that are critical. So when forests are cut and the soil dries out, it is worse than when surface waters, such as rivers, dry up. Soil water evaporates very quickly with annual crops and annual cultivation. The solution is to plant perennial tree crops to shade soil and hold water in soil.

Most of the water associated with rivers runs underground. To keep these lines flowing, the hills and ridges must be kept in permanent forest. The underground waters link up with aquifers and wetlands, lakes and springs. They are a little-known landscape.

Rehabilitating rivers, creeks, streams and wetlands

We must rehabilitate and maintain in good condition all the surface waters that we have abused. This is achieved by keeping the edges of all surfaces under permanent vegetation and shade, which carry out the following functions:

- trap silt and run-off
- take up toxins and excess materials in the water (for example, fertilisers and biocides) as they run off our land, gardens, streets and industry
- hold the edges against erosion
- absorb the energy of floodwaters
- provide breeding grounds and protection for indigenous animals, both terrestrial and aquatic
- offer cooler water temperatures through shading.

We can start rehabilitation by removing dams on rivers and creeks and drainage from paddocks to let wetlands re-form. If riverbanks have been badly incised by erosion, begin by fencing the whole area off (see Figure 4.12). Place the fence boundary a minimum of 30 metres each side of the riverbank. Replant with a variety of indigenous plants and include reeds, shrubs and trees. Don't worry too much about weeds because if you plant densely they will be shaded out quite quickly, or you can remove them later. Around swamps and wetlands, water plants of the reed family used for water cleansing are essential.

If you have an incline or fall in the creek or riverbed and want to reduce the speed of the water, slow it down by weaving barriers from local indigenous shrubs—in my case this is tea-tree and hakea with seedpods on the branches—and place these across the flow at about 10-metre intervals (see Figure 4.13). These barriers will slow down the water, trap waste and send indigenous seed down the river to grow further downstream. Do not use straw as it composts, grows weeds and is bad for water quality.

Within a few years your creeks and rivers will be well vegetated, cool and damp. Ephemeral waters will run longer and cleaner after rain.

Record your work in your observation notebook because your results will be useful to others in your community. The increase in plants and animals and change in water quality will give you great joy.

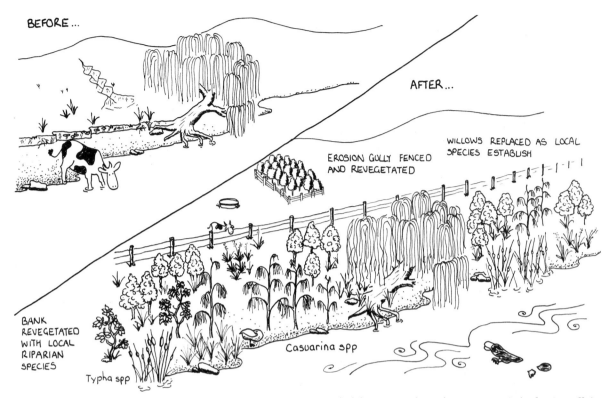

BEFORE...

AFTER...

EROSION GULLY FENCED AND REVEGETATED

WILLOWS REPLACED AS LOCAL SPECIES ESTABLISH

BANK REVEGETATED WITH LOCAL RIPARIAN SPECIES

Typha spp

Casuarina spp

Figure 4.12 Rehabilitating creeks and streams. Begin by fencing off the area to prevent further erosion and replanting with indigenous plants.

Bores and ground water

Ground water is any water below the ground surface. It includes both water percolating down to the water table and standing water below the water table, and plays a huge role in what happens at the surface. Ground water is constantly recharged from precipitation and discharged through waterfalls, soaks and springs, and slowly feeds creeks and rivers as clean water. Whole ecosystems are defined by it. Salts dissolved in ground water are maintained at reasonable levels until the trees are cut down and then it rises and concentrates the salts at the surface and salinity occurs. Ground water must be kept clean. Caring for it would mean replacing water-hungry irrigated crops such as cotton with perennial income-earning crops that take less water.

When ground water is extracted by bores and wells, the natural dynamic balance is disturbed and the environmental consequences may be extremely unacceptable. There is little knowledge of the economic benefits of ground water extraction compared with leaving ground water in the ground,

because we don't know the amount of recharge needed to sustain healthy ecosystems. As rainfall is declining, ground water recharge will be less. We have little understanding of how climate change threatens ground water resources. However, ground water and surface water are completely interdependent and reduction of one will rebound on the other.

The environment is a legitimate user of ground water and we must not steal from it. Many ecosystems carry out valuable functions and are dependent on ground water for their survival. As a corollary, if ground water is depleted, ecosystems can die. To maintain the integrity of ground water it needs:

- protection from unregulated bores and pumps
- recharge areas, which are the permanently timbered tops of hills that enable 5–10 per cent of rainwater to seep through to recharge the ground water.

Every farm design needs to demonstrate designated permanent recharge protection zones

1. CROSS-SECTION (FROM FRONT)

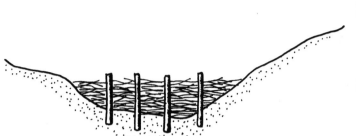

1. KEY BRUSH 300mm INTO BANK AND 150mm INTO CREEK BED

2. BRUSH - USE LOCAL CREEK SPECIES WITH SEED CAPSULES

2. PLAN VIEW (FROM ABOVE)

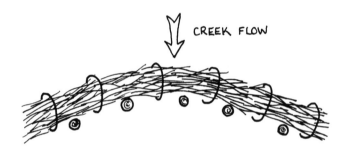

CREEK FLOW

3. POSTS ON DOWN-SIDE OF BRUSH BUNDLES

4. WEIR V-ANGLE UP-STREAM TO SLOW DOWN AND DIRECT FORCE OF WATER TO SIDES

3. PROFILE VIEW (FROM SIDE)

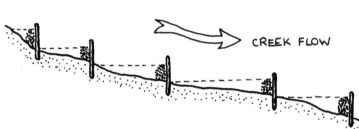

CREEK FLOW

5. SEVERAL WEIRS ARE BUILT ALONG THE CREEK.

BOTTOM OF UPPER WEIR IS THE SAME HEIGHT AS TOP OF NEXT LOWER WEIR

Figure 4.13 V-notch weir design for creeks slows water velocity, promotes sedimentation, provides indigenous mulch and reseeds downstream.

on hills, and carefully monitored and seldom-used bores and wells.

Lakes and ponds

Lakes and ponds are closed systems and, like ground water, are often difficult to flush out, so it is important to maintain their water purity. The first rule of managing them is to keep a permanent edge of reeds around the entire water body, and to have it fenced beyond that if the pond is a source of drinking water. It is important that animals such as pigs, ducks and dogs cannot reach the water because all these animals can carry diseases transmissible to humans.

Water entering ponds and lakes needs to be pure and free of shampoos, detergents or soaps

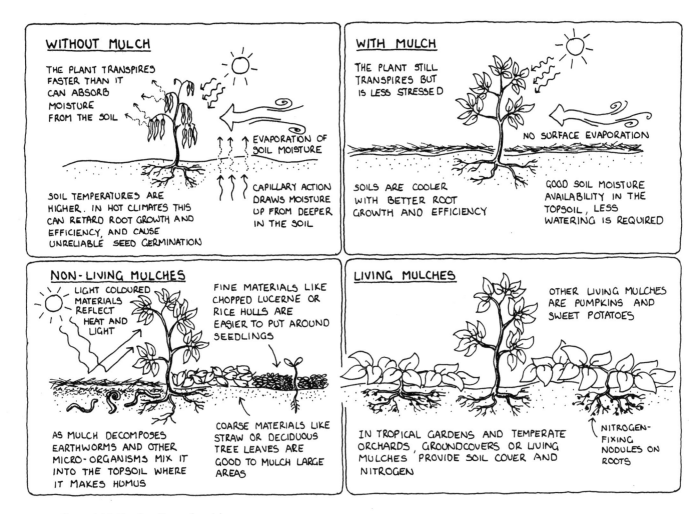

WITHOUT MULCH

THE PLANT TRANSPIRES FASTER THAN IT CAN ABSORB MOISTURE FROM THE SOIL

EVAPORATION OF SOIL MOISTURE

SOIL TEMPERATURES ARE HIGHER. IN HOT CLIMATES THIS CAN RETARD ROOT GROWTH AND EFFICIENCY, AND CAUSE UNRELIABLE SEED GERMINATION

CAPILLARY ACTION DRAWS MOISTURE UP FROM DEEPER IN THE SOIL

WITH MULCH

THE PLANT STILL TRANSPIRES BUT IS LESS STRESSED

NO SURFACE EVAPORATION

SOILS ARE COOLER WITH BETTER ROOT GROWTH AND EFFICIENCY

GOOD SOIL MOISTURE AVAILABILITY IN THE TOPSOIL, LESS WATERING IS REQUIRED

NON-LIVING MULCHES

LIGHT COLOURED MATERIALS REFLECT HEAT AND LIGHT

FINE MATERIALS LIKE CHOPPED LUCERNE OR RICE HULLS ARE EASIER TO PUT AROUND SEEDLINGS

AS MULCH DECOMPOSES EARTHWORMS AND OTHER MICRO-ORGANISMS MIX IT INTO THE TOPSOIL WHERE IT MAKES HUMUS

COARSE MATERIALS LIKE STRAW OR DECIDUOUS TREE LEAVES ARE GOOD TO MULCH LARGE AREAS

LIVING MULCHES

OTHER LIVING MULCHES ARE PUMPKINS AND SWEET POTATOES

IN TROPICAL GARDENS AND TEMPERATE ORCHARDS, GROUNDCOVERS OR LIVING MULCHES PROVIDE SOIL COVER AND NITROGEN

NITROGEN-FIXING NODULES ON ROOTS

Figure 4.14 The functions of mulch.

unless you have very special products that do not contain phosphates or laurel stearate. In warmer climates the edge planting can be of productive plants such as lemon grass, lotus and water chestnuts. Water plants such as lotus, hyacinth and waterlilies covering 30 per cent of the surface area keep the water clean and cool. Note, however, that in Australia water hyacinth is a major water weed of rivers and should not be used. These water plants can be harvested for mulch or pig food.

Design techniques for dry land and drought in gardens and orchards

The principle behind techniques for caring for dry land in times of drought is to deliver the right amount of water to the plant and as close to its root system as possible. Spread mulches thickly.

Try these:

1. Put all the sources of incoming water on your site plan. Show the pipes bringing water in. Show roofs from which you could capture water.
2. Create a table in your workbook of your water consumption. This table is called your water audit and it is an analysis. Set yourself goals to reduce the amount of water you use.
3. Look at your land and identify 'run-off' areas, like driveways, roads, sheds and house roofs. Notice where they shed water; these are called 'run-on' areas. Now design water-capture systems for the soil and in biomass, so the water will stay where you want it.
4. Draw up specific techniques for repairing wetlands (including small undulations in your land) and rehabilitating streams and rivers.
5. Work out how to store water in land storage systems, even on very small land, so you will have water available during dry periods.

CHAPTER 5

Climates and microclimates

For the last 10,000 years or so Earth's climate has been relatively stable and predictable. And that predictability is important for most facets of our lives. Now, there is widespread agreement that human activities over the last two centuries have destabilised world climates.

One of the latest conferences declared, rather depressingly, that 'there is now no stopping global warming' (ABC Radio, March 2005). In Russia, an area of permafrost the size of France and Germany that has been frozen for 10,000 years has started to melt and will have a huge impact on further destabilising climates. The effect of global warming will be that every continent will find its climates unreliable and often unpredictable with extremes of heat, drought, storms, cyclones and ocean rise.

The *Sydney Morning Herald* of 9 September 2005 reported an English study which said that 'present forecasts of climate change could be seriously under-estimated because of huge amounts of carbon pouring out of the earth ... there was little that could be done to tackle the problem without addressing the fundamental question of human carbon dioxide emissions. If we were prepared to turn the whole of arable England back to trees that would work but it's not practical.' Basically, we have over-cleared land everywhere.

In permaculture we apply the precautionary principle, which you read about in Chapter 3 (see page 19), to design settlements and enterprises that are robust and have a chance of withstanding stress and disasters. To do this we work with the elements of climate in siting buildings and growing areas, and identifying sites that need to be enhanced or

protected. You started this process when you carried out your inventory and began your observations.

The concept of climate is too big and ill-defined to use effectively in a site analysis and design. Instead, we consider climate by identifying each of its three major components and then studying them separately to see how they are destructive or beneficial on a specific site, and how they interact and follow cyclic patterns. The three components are precipitation, wind and radiation (sun). Together they act on the huge continental landmasses and ocean currents.

 Our ethical task is to:
- ensure climatic stability
- reduce atmospheric pollution.

 Our design aims when working with elements of climate are to:
- modify extremes of climate
- reduce the risk of animal and crop failures
- choose appropriate animals, plants, buildings
- select correct elements of climate to work with—for example, for solar or wind energy
- achieve more energy-, materials- and water-efficient buildings

- endure, withstand or avoid droughts, floods and other disasters.

If we don't have design aims:
- we contribute to climate destabilisation and disturb microclimates
- we hasten the breakdown of ecosystems
- we lose opportunities for diversity of niches, edges and species.

Precipitation

Precipitation is rain, snow, fog, sleet, hail, mist and frost. All precipitation is seasonal and follows fairly predictable patterns. Cold winter rain comes from the South Pole in the southern hemisphere and the North Pole in the northern hemisphere and always travels from west to east in both hemispheres.

Hail, snowstorms, fog and frost also have patterns of time and place. For example, frost forms on the ground on very cold, cloudless, still nights but it doesn't form under trees, eaves of houses or when there is a breeze. Fogs drift in from oceans and over mountains predictably in autumn and in some places in winter. Cyclones follow patterns in tropical climates, usually on east coasts of continents. Long-term residents often have sayings based on climate patterns. Our local one here is, 'It can frost up until November 15.'

By understanding patterns of precipitation you can plan your growing year to take advantage of favourable conditions or minimise the impact of unfavourable conditions. For example, it is useful to predict likely rainfall or drought for crop planting and water harvesting and storing. However, with the increasingly erratic nature of climates worldwide, it is safer to act as if there will be droughts, floods and other natural disasters and design sturdy environments and enterprises.

Different forms of precipitation occur because water changes form. When it changes from ice to liquid to gas the process is called evaporation and is accompanied by cooling; when it changes from gas to liquid to ice the process is called condensation and there is warming.

As permaculture designers we use knowledge of these two processes to:
- design structures with good temperature control
- select appropriate renewable technologies for heating and cooling
- retain water in the soil
- position plantings and structures appropriately.

Wind

Winds are caused by the earth's rotation and the differential heating of land and sea surfaces. For example, deserts radiate intense heat upwards and this sucks in cooler, moist air from the oceans, bringing rain as it moves in. When this happens for an extended period on a vast scale, monsoon climates are created. When it occurs daily, the effect is diurnal and local. Winds are divided into orders. The large orders such as monsoons, cyclones and typhoons are all part of climate, whereas small breezes that move up and down hills daily or the canyon effects in cities are part of the local microclimates. Whether winds are of large or small orders they all have patterns.

Like precipitation, world climates have fairly predictable wind patterns, although these too are changing with global warming. Knowing these patterns helps you to:
- harvest wind energy for electricity
- design your home and animal shelters to benefit from, or minimise, the impact of winds
- design and plant protective windbreaks
- select wind-tolerant species
- change activities for different seasons
- build buffers against climate change.

Radiation

Most radiation is white light. It comes from the sun and is absorbed by water, plants, soils, materials and animals, and is changed into other energy forms such as heat and chemical energy. It can be:
- radiated back as heat energy from the soil, water and some materials

49

- turned into chemical energy by green plants when they photosynthesise
- turned from light energy into heat energy after passing through glass.

Photosynthesis is the process by which plants turn light into food (chemical energy). Plants absorb different wavelengths in different parts of their structures—for example, the back and front of their leaves. Young leaves absorb different amounts of light to older leaves. In addition, in the tropics, dark green and red leaves absorb large amounts of radiation and assist in cooling the environment, while in temperate climates, where plants have light-coloured leaves and bark, light is reflected back to increase light and its absorption in the environment. Light is vitally important to life.

- Light absorbed by plants in photosynthesis has a cooling effect. One impact of removing forests and other vegetation is to increase heat and light in the Earth's biosphere, causing thermal pollution—global warming.

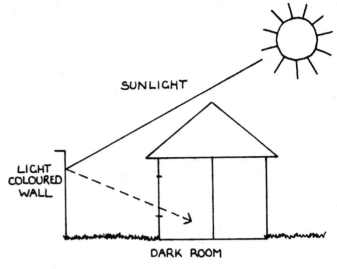

Figure 5.2 *Reflection of light. A light-coloured wall can be placed so that is faces the sun and reflects light into dark rooms on the shady side of the house.*

So a city parking lot or a piece of bare ground is hotter and glarier than a similar vegetated area. Soil moisture, which helps to stabilise climate, is also lost.

- Certain seeds have evolved mechanisms so that light triggers germination. They are often dry-climate or desert plants. Others need short days to germinate—that is, long periods of darkness. This is called photoperiodism and tends to occur in high latitudes. With plants in equatorial regions, where day and night are about equal, light has less effect and instead it is usually rainfall that triggers germination. The same mechanisms occur with blossom set and seed set. Light also initiates oestrus in some animals.
- Light is absorbed and stored by dark bodies, also called thermal mass, which later radiate it back as heat. Figure 5.1 shows how this process can be used to warm a room.
- In contrast, light is reflected from light-coloured bodies such as plants, water and materials, and the reflected light can be used by plants and animals. Figure 5.2 uses reflection of white light to increase light in a room.
- Light is also turned into heat energy after passing through glass.
- Finally, light changes the habit (shape) of plants. Where there is little light, plants grow

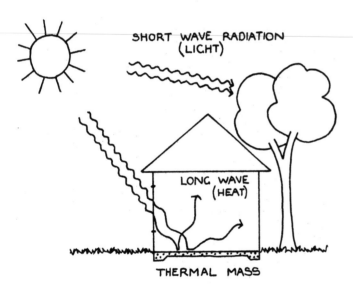

Figure 5.1 *Use of thermal mass for heating. Light from the sun is absorbed by surfaces and is converted into heat energy which is re-radiated into the surroundings. This process can be used to capture and store energy for heating your home. If you expose a concrete slab floor to sunlight during the day, light energy is absorbed and converted to heat which is conducted throughout the slab and re-radiated later when the surrounding air has cooled. This helps to warm the room at night and generally maintain a more even temperature in the house.*

tall and straggly in an effort to capture more light. Where there is full light plants tend to become round. It is important to know this when planning plant spacing in gardens and forests.

Interaction of precipitation, wind and radiation

Each of these components of climate interacts with the others all the time and changes in impact and force. On any site, however, one may be more useful or destructive than another. For example, in some latitudes and seasons, winds are destructive and limiting. If you know their patterns you can modify the site you are designing. For building structures, knowledge of radiation as light energy and its forms can help you to retrofit a building to modify the extremes of heat and cold.

However, the focus of most of your work will be the microclimate, because it is here you can achieve the greatest gain and impact for the least work. In Part Three you will analyse the components, or elements, of microclimate and then put them together in a microclimate study of a home site you have selected to study.

Microclimates

The place where you live is a microclimate of the larger general climate. In my area and climate, the cold weather always comes from the south. But at my place, the cold winds and rain always come from the west. I live on a ridge and two large valleys channel the southerly winds around to the west Also, data shows the sun rises at 6.30 in autumn for my latitude and longitude but at my place it rises at 8.30 because of a hill to the east of my house. These special effects comprise part of my microclimate.

In our look at climate, we examined three elements: precipitation, wind and radiation. The major climates always have local variations in temperature, wind speed, direction, relative humidity and light levels. This is because the set of five local microclimate factors acts on the larger climate to give rise to a series of microclimates nested within

the larger climate. These local factors are:
- topography
- soils
- water masses or bodies
- artificial structures
- vegetation.

Overall, in designing diverse and stable landscapes, knowledge of microclimate is more important than knowledge of general climate. Microclimates can be a rich source of diversity, and so your design should take advantage of those occurring naturally rather than eliminate them. You can increase the diversity of plants and animals in your site by applying your knowledge of the elements of climate and microclimate through design. For example, in a cool temperate climate you may be able to grow an almond tree if you place it against a warm dry westerly wall, and you can create special niches for more plants by using stone paths to give a warm root run.

Characteristics of microclimates

The general climate sets the broad limiting factors of your site and knowing and analysing microclimates will help you to modify these. The factors of microclimate, like those of climate, are dynamic and interactive. Take any site and start your observations. For example, steep slopes usually have shallow soils at the top and deeper soils at the base, are dry at the top and damp at the base.

Observation is only the beginning, however; you need to connect your observations and make deductions for design strategies and techniques. This is called microclimate analysis. It enables you to:
- read the landscape and predict microclimate effects—you may notice that paint is peeling off one side of the house and this shows the direction of the prevailing winds or drying winds across your site
- modify climate extremes—you can terrace slopes to capture more sunlight and warmth or to hold water on a dry steep hillside
- design effective strategies to achieve the microclimates you require—the careful placement of windbreaks as suntraps can increase temperature

SHADED ASPECT

SUNNY ASPECT

Figure 5.3 Aspect and shading. On the sunny side of a slope, shadows cast by objects are significantly shorter than the shadows cast on the shaded side.

- extend the growing season and biodiversity —frosts can be avoided by the use of vegetation and structures or directed breezes
- live more comfortably and use less of non-renewable resources—design an efficient solar home.

Topography

For our purposes the main elements of topography are aspect and slope.

Aspect

Aspect is the direction that a slope faces and is characterised by the amount of radiation it receives. Figure 5.3 shows how the aspect of a slope affects warming, cooling, and shadows cast by structures and vegetation. On the shaded side of the slope shadows can be up to three times longer than on the sunny side.

Aspect gives rise to thermal zones and/or cold sinks. These occur because:

- air moves faster uphill than downhill
- cool air is heavy air and moves downhill
- warm air is lighter and moves uphill
- cool air replaces warm air by pushing it up or sliding underneath it.

When cool air flowing downhill is impeded by any barrier, it pools and is called a 'cold sink'. On slopes warm air is pushed upwards by cool air. When this warm air is trapped by a building or plants, it is called a 'thermal zone'. If there is no barrier, the warm air will continue to drift up the

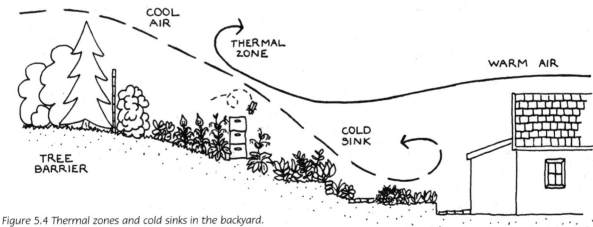

COOL AIR

THERMAL ZONE

WARM AIR

COLD SINK

TREE BARRIER

Figure 5.4 Thermal zones and cold sinks in the backyard.
As the warm air rises it can be trapped by fences and trees to create a warm thermal zone on the upper slope. The downhill area will be cooler and more prone to frosts as the cool air sinks at the end of the day. The same process occurs on a much larger scale in valleys.

slope until it cools. You can design thermal zones and cold sinks to suit your needs for more warmth or coolness, or to site a home or suit a crop.

Aspects give rise to winds with differing qualities. Winds tend to be warmer and drier on western slopes and flow upwards because the west receives more intense radiation after midday. On the aspect away from the sun, cool heavy air drifts down and if it is blocked will then form a very cold little microclimate. Cooler and wetter winds arise from poleward slopes, while winds from the east tend to be reasonably moderate and pleasant on eastern coastlands, and evening winds on west coasts come off the sea and are refreshing. This knowledge, verified by observation, helps you in siting and orienting homes and animal housing for the maximum realisation of renewable energy and site potential.

Aspect also affects your choice of plant and animal species. Some species prefer the eastern aspect with the morning sun, and others prefer the warmer, drier western slopes.

Slope

Slope affects wind speed because the steeper the slope, the faster wind moves uphill. This has implications for managing wildfires, capturing wind energy, and siting windbreaks (see Figure 5.5).

Slope also has a major effect on water speed because water increases its velocity as it moves downhill. Sloping land erodes faster and more severely than flat land. Fast-moving water is usually very destructive; however, it can be harnessed for hydropower, or controlled and redistributed, or erosion-prevention works can be effectively designed. People living in hot, wet, mountainous areas of the world terrace their slopes to prevent water erosion.

Slope affects cultivation techniques. Because of the destructive nature of cultivation machinery, and the tendency of slopes to collapse in landslides, it is a good general rule that slopes of greater than 15 degrees from the horizontal are better placed under permanent productive trees.

Soils

Soil used to be considered the least important factor in determining microclimate but we now know that the moisture in soils has a major impact on climate stability. Soil texture and structure affect the absorption, shedding or evaporation of water. Soil types give rise to specific local effects and affect what can be grown.

Clay soils hold more water, shrink and swell when drying and wetting, and respond differently to different cultivation techniques. Clay soils crack open when dry and are ready for the rain to penetrate deeply in the first storms. Sandy soils drain fast, don't shrink or swell, and are easy for cultivation machinery.

Soils like to be covered. Bare soils reflect more heat and light compared to covered soils. They are also more vulnerable to wind and water erosion and desiccation.

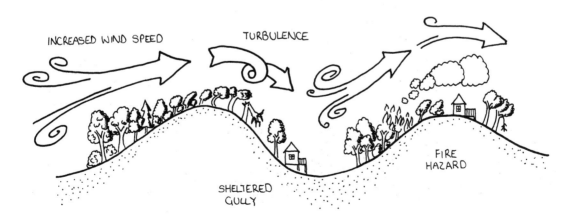

Figure 5.5 Slope and wind speed.

Water bodies

Rivers, lakes, dams and ponds modify climate and generally contribute to more pleasant microclimates. This is because water gains and loses heat more slowly than land. Water bodies provide the following benefits in the microclimate:

- Because they reflect light and warmth they can be situated so as to warm buildings. A wider range of plants can be grown around lakes and dams. Sites around lakes and rivers can be 5°C warmer than land with no water bodies.
- Sites close to oceans and seas have a 'maritime' effect, where cooling evening breezes relieve heatwaves. Inland climates have 'continental' effects, with extremes of heat and cold occurring in a single day because land both gains and loses heat fast.
- They increase humidity in the air and extend your choice of species—many palms require high levels of moisture or humidity in the air.
- They provide habitats for water-loving plants and animals, and they add immeasurably to pest control because so many predatory animals need regular access to water.
- They modify temperature extremes because water bodies cool warm air and also warm cool air. After heavy summer rains, when soils hold a lot of water, the wet soil can act like a local lake to modify climate. Weather forecasters often predict warmer winters after wet summers.

Artificial structures

Structures that affect microclimates range from dog kennels and duck houses to multi-storey buildings, and include items such as fences and roadways. For example, structures with steep 'slopes', such as walls and roofs, can contribute enormously to environmental damage through water run-off or excessive reflection of heat. On the other hand, structures can also be beneficial when used to:

- trap and store water
- collect and store light as heat
- grow plants in, on and around, and so add vertical space to small areas
- funnel or reduce winds
- increase the growing season by providing thermal mass

Figure 5.6 Vegetation modifies the environment.

MICROCLIMATE FACTORS	1. BACK FENCE	2. SIDE FENCE ON WEST	3. EASTERN BOUNDARY	4. COURTYARD	5. FRONT YARD
TOPOGRAPHY • ASPECT (SUN) • SLOPE (WIND)	SOUTHERLY ASPECT, NO WINTER SUN LEVEL; EXPOSED TO SOUTH	EASTERLY ASPECT, GOOD MORNING SUN SLOPE INCREASES UPWARDS	WESTERLY ASPECT, GOOD AFTERNOON SUN SLOPES UPWARDS	NORTHERLY ASPECT, SUNTRAP LEVEL; COLD SINK	SOUTHERLY ASPECT, NO WINTER SUN EXPOSED TO SOUTHERLY WINDS
SOIL • COVER • COLOUR • TEXTURE • MOISTURE	COVERED BY WEEDS DARK; WITH ORGANIC MATTER COARSE ALWAYS DAMP	COVERED BY GRASS LIGHT; NO ORGANIC MATTER SANDY DRY	GRASS AND WEEDS LIGHT; NO ORGANIC MATTER CLAY AND SAND FILL DRY	RAISED GARDEN BEDS DARK; WITH ORGANIC MATTER COMPACTED CLAY/SAND ALWAYS DAMP AND SOUR	COVERED BY GRASS DARK; WITH ORGANIC MATTER SANDY LOAM DAMP AND SOUR
VEGETATION • BARE • GRASS • SHRUBS • TREES	BLACKBERRY AND WEEDS ONE SMALL EUCALYPT	GRASSED	GRASS AND WEEDS	MOSTLY PAVED WEEDS AND GRASS IN BEDS	GRASS ORNAMENTAL SHRUBS
WATER • RUN-OFF	NO RUN-OFF; POORLY DRAINED	RUN-OFF DOWNSLOPE	RUN-OFF DOWNSLOPE	NO DRAINAGE; WATER SITS IN POOLS	RUN-OFF DOWNSLOPE
STRUCTURES • WINDBREAK • WINDFUNNEL • COLOUR (THERMAL MASS) • INCREASES HEAT • INCREASES COLD	SIDE FENCE AS WINDBREAK HEAT AND LIGHT REFLECTED BACK FENCE CREATES SHADE	SIDE FENCE AS WINDBREAK HEAT AND LIGHT REFLECTED SIDE FENCE CREATES AFTERNOON SHADE	WIND FUNNEL BETWEEN HOUSES LIGHT, CONCRETE WALL REFLECTS AND HOLDS HEAT	HOUSE ACTS AS WINDBREAK PAVING REFLECTS LIGHT AND HOLDS HEAT HOUSE TRAPS COLD AIR IN WINTER	WIND FUNNELLED UP STREET HOUSE SHADES GARDEN IN WINTER

Figure 5.7 Microclimate study of Rob's place.

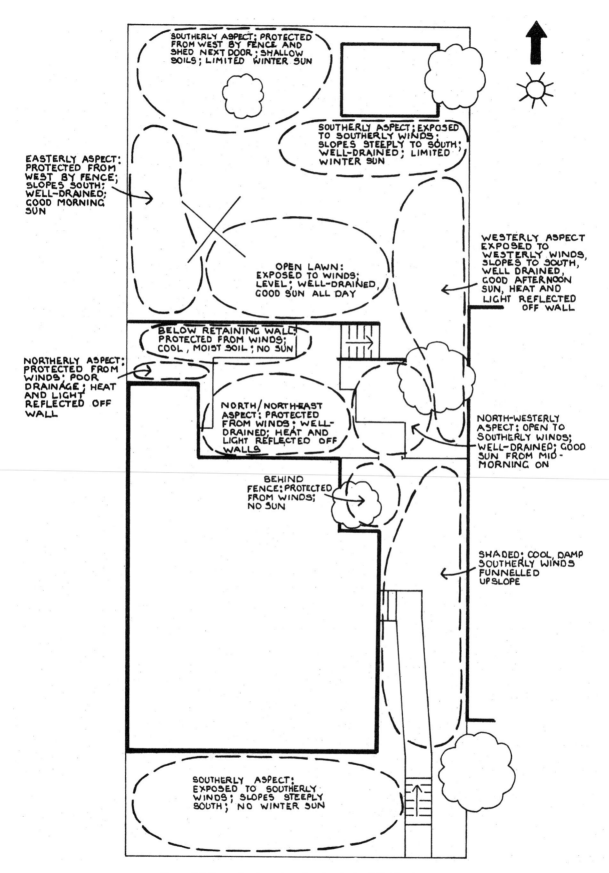

Figure 5.8 Example of a microclimate study at Rob's place.

- ripen plants by reflecting light
- drain land by using contour banks or mounds and ditches
- reduce noise pollution.

Vegetation

Vegetation interacts with and changes other microclimate factors such as soils and water.

- It absorbs heat and light. Without vegetation, solar radiation and reflection is very intense; the soil becomes vulnerable to drying out and losing nutrients, and is exposed to erosive forces.
- Vegetation acts as a carbon sink. It is considered to be the most effective way—and possibly the cheapest—to mop up surplus atmospheric carbon dioxide, which is such a large contributor to global warming.
- It provides habitat, windbreaks, suntraps, shelterbelts, firebelts and firebreaks.
- Vegetation regulates soil temperature, keeping it warmer in winter and cooler in summer than bare ground.
- Dust, diseases and excessive moisture from winds are all filtered by vegetation, as are pollutants from soil and water.

Vegetation has other characteristics that you can incorporate in your design (see Figure 5.6):

- It is adapted to suit its climate of origin, for example rainforest plants often have large, dark leaves that absorb much heat and light and release water vapour, making a microclimate cooler than it would have been otherwise.
- It is used as an architectural tool. In 'biotechture', plants' traits are used to provide desired microclimates.

 Try these:

1. Find the climate figures for your area from the Meteorological Bureau. Try to match temperature with rainfall and evaporation. What did you find out?

2. Do a microclimate analysis of a site. Copy your base plan from the last chapter or place a sheet of tracing paper over it. Using the microclimate factors, mark out all the microclimates of the site on the plan. Look at Rob's microclimate analysis in figures 5.7 and 5.8 to see how he did it.
   - Where do the main summer and winter winds cross your land? Verify this from your own experience.
   - Decide whether wind or precipitation is the most damaging, and in what form. For example, determine whether it's heavy frosts in September, or very strong winds in July. Be as precise as you can.

3. Identify two microclimates and analyse them for the factors you have just read about—that is, topography, soil, water bodies, structures and vegetation, and how each functions and interacts with the others.

4. Find these microclimates:
   - a sunny spot where you like to breakfast on a cold and windy morning
   - the most weather-damaged aspect of your house
   - shady places in the garden that get little or no sun
   - cool places to be when the weather is abnormally hot.

5. Observe and record your answers to the following questions:
   - Which insects are active in summer compared to winter?
   - Which plants flower and fruit as the days grow longer?
   - Which plants fruit as the days grow shorter?
   - Which plants turn yellow after many cloudy days, and which ones are not affected?

CHAPTER 6

Soils: living organisms

All good gardeners and farmers have a passion for soils. They pick them up, run them through their fingers and smell them. And as you talk to these people you will find they tell you how they feed and care for their soils.

They will tell you what the soil was like when they first started gardening and, if you have time, they will take you to a place where the soil is thin and lifeless to actually show you how it once was. As you continue your relationship with the earth, you too will find yourself acquiring these same convictions. A friend of mine grows lyrical as she tells me about the stash of horse manure that she has discovered to feed her soil.

It is highly likely that neither you nor your children will see a healthy soil in your lifetime. Soils are not respected. They are compacted, eroded, dumped on, spat on, moved, covered, cleared, levelled, poisoned, flooded, drained, mined, turned upside down, and fertilised and sprayed with chemicals. Our ethical task is to improve those soils that will give high productivity and have been severely damaged, and to leave alone and respect those natural soils that support special ecosystems such as swamps, deserts, coasts and mountains. Some ecologists say that the greatest task on Earth is to restore soils to a healthy state and that we ignore this at our peril.

It is useful to think of soils as living organisms. Like forests, plants, water and climate, which are highly integrated living systems, they defy precise measurement and their functions are complex and synergistic. Soils are an unknown landscape that we barely understand.

 ## Our ethical task is to:
- respect and leave untouched all naturally occurring soils that support unique ecosystems
- repair and protect all damaged soil
- respect soils as living organisms.

 ## Our design aims for soils are to:
- carry out a whole site soil analysis
- recognise and repair damaged soils
- choose and use nutrients strategically.

 ## If we don't have design plans for soils:
- the soil can become sterile
- acidity problems from poor fertiliser and water application techniques can result
- destructive farming techniques will continue to be practised
- lakes, rivers and ground water can be polluted and contaminated
- unhealthy food with too many nutrients or toxins in it will be produced
- soils will be exposed to wind and water erosion
- desertification and desiccation may result
- all soil life may be killed
- soil, crops and water will be lost.

Ecological functions of healthy soils

A healthy soil breathes, recycles waste, promotes active growth, stores nutrients and cleans water. Soils enable basic life processes for all living things. When soils are damaged, so is life as we know it.

Soil type and quality

Although very important, soil is not the primary selection factor for land because there are effective and proven techniques to repair and build soils quickly. However, if you are fortunate enough to have land with rich, sticky, red basalt soils then you are saved many years of hard work.

You will find that each part of your land, depending on its microclimate and use, has different soils. So respect that and work with it.

A simple analysis of your soil

Take three clear plastic or glass jars. Collect three different soil samples and put one in each jar— about 25–30 per cent of the jar's capacity. Now add water to about 80 per cent of the volume of the jars.

Shake each jar very well. Leave to settle for 24 hours then place the three jars in a row and look for the following:

- Loose, unbroken organic materials floating on the top. These indicate the soil has a nutrient bank to break down.
- Clear or murky water. Murky water has dissolved or suspended nutrients in it, which are immediately available to the plants; it's good.
- A fine silt layer on top of a denser one and then finally some coarse sand or gravel at the bottom. These tell you about your soil particle fractions. A reasonable layer of silt is also good because it, too, is a nutrient. A proportion of sand tells you your soil will drain well.
- Look at the colours of each sample. The closer to red or black the better.

Keep the samples and mark on them where they came from and the date. Over the course of one year, work on one of these soil sites, then repeat this test. Have you changed the soil?

Composition of a healthy soil

A healthy productive soil has a balance of:

- moisture
- gases
- mineral fractions
- micro-organisms
- organic matter.

Together these contribute to the main function of soil, which is to break down large physical and organic compounds to simpler ones that are absorbed by plant roots and used by soil organisms. Each of these five main elements not only differs depending on the soil but also interacts with all the other elements.

TABLE 6.1: FUNCTIONS OF HEALTHY SOILS

Function	How it works
Cleansing	Absorbs and filters some of the toxins in organic matter and transforms them to less toxic substances (for example, nitrites to nitrates).
Holding/support	Provides a medium to secure root systems of plants and support structures.
Respiratory	Absorbs atmospheric gases and recycles them from soil life via the metabolism of roots, organisms and the atmosphere. Think of soils as large digesters.
Digestive	Breaks down large physical and organic compounds (often wastes) to simpler ones, which can be absorbed by plant roots and used by other soil organisms as nutrients.
Storage/bank	Absorbs and holds water and nutrients for future use by plants.
Solvent	Dissolves natural chemicals for the roots to take up.

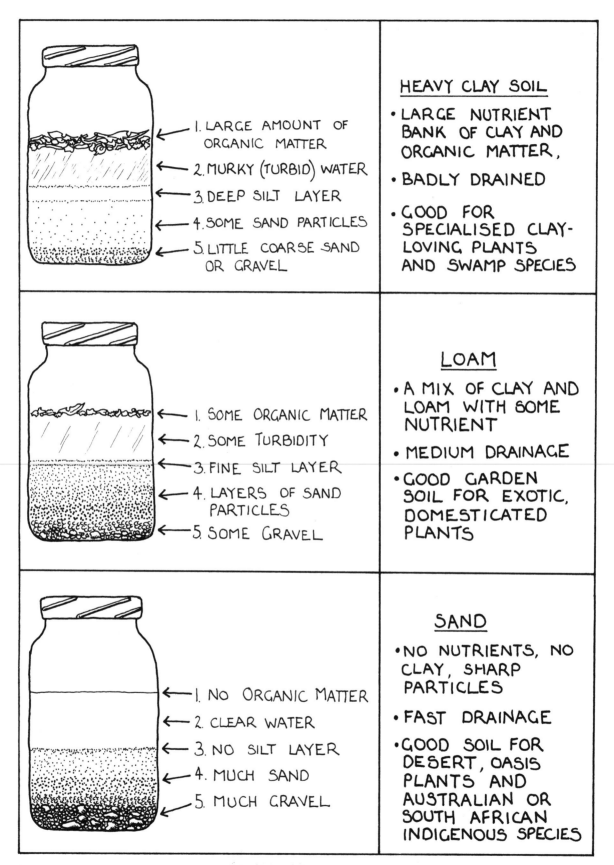

Figure 6.1 A simple soil analysis.

Moisture in soils

Water in soil is a weak acid or alkaline solution carrying the soluble nutrients that plants absorb through their root systems. And while these liquids must drain or the soil becomes waterlogged, they must not drain too fast or the soil dries out quickly. All plant nutrients exist in water-soluble forms.

The pH of soils refers to the acidity or alkalinity, which in soils is a measure of the solubility of various nutrients. So, if the soil has a pH of 8.0 then some nutrients are soluble and the soil is said to be alkaline. If the pH is 5.0 then other elements are soluble and the soil is said to be acidic. Adding lime to an acid soil to make it more alkaline, or sulphur to an alkaline soil to make it more acidic, changes soil pH. Most of the world's plants grow within a pH range of 5.5 to 8.0, and within this some plants will struggle and others thrive. Figure 6.2 shows you the acid–alkaline tolerances of different plants.

It is rarely desirable to apply a single chemical to change pH however, because we don't know how it interacts with all the other soil variables. The best solution for soil problems is always the addition of organic material.

Moisture moves from gas to liquid depending on air pressure and temperature. In soils, moisture as a gas moves upwards due to evaporation from wind or sun, or the pull of the water from the roots to the leaves due to transpiration. Water also moves laterally along bedrock. Gases and liquids move downwards after rain. Over-watering leads to leaching, which is the opposite of salination. Nutrient salts are washed down out of the root zone into the deep layers of the soil, even the aquifer. Some tree root systems penetrate deeply looking for this water.

In cloudy places where there are mists and fogs but no recorded rain, this moisture is absorbed by plant leaves and diffuses into the soil.

Scientists have recently declared that soil and vegetation moisture are extremely important in maintaining climate stability. Soils must not dry out due to poor farming methods and vegetation removal. In permaculture, keeping moisture in soils is important for plant growth. Refer to Chapter 4 on water for some techniques on how to do this.

Gases in soils

Gases in soils change pressure and type at different times of the day and in different seasons. They also move into and out of soils. How freely they move depends to some extent on the texture and structure of the soil. Gases in air are exchanged with those given off by plant roots and soil micro-organisms. If there is adequate oxygen then soil tends to be sweet-smelling. If there is little oxygen then other gases such as sulphur dioxide build up and soils smell rotten.

It has recently been found that ethylene gas is particularly important because when it is given off organic matter is broken down. (Ethylene gas is known as the 'ripening' gas and is given off when bananas, oranges and other fruit ripen.) In soils it cycles with oxygen, increasing the build-up of micro-organisms and other soil materials. Techniques that let more oxygen into the soil, such as forking or deep ripping without turning over the sod, are soil improvers because they assist the ethylene cycle.

Mineral fractions in soils

This refers to the type and size of rock and clay particles in soil. The 'feel' of a soil when you rub it between your finger and thumb defines its texture. When the particles are mainly sand and coarse gravel then soil feels rough and its texture is said to be gritty. When soil has minute particles, usually of clay minerals, then the soil has a smooth feel and is said to be silky (see Figure 6.3).

Gritty soils:
- drain quickly, and dry out quickly
- have few soil fungal diseases
- leach out soil nutrients (wash them into lower soil layers).

Silky soils:
- drain slowly and hold water for a longer time
- shrink when dry and swell when wet
- hold soil nutrients on the surface of clay particles
- when bare, form a claypan—a concrete-like surface.

If a soil is almost pure clay or pure sand it is called a 'difficult' soil. In both cases soil texture and

ACID AND ALKALINE TOLERANCE

QUITE ACID 4.0 – 6.0	SLIGHTLY ACID 6.0 – 7.0	NEUTRAL TO ALKALINE 7.0 – 7.5
BLACKBERRY	APPLE	ALFALFA
BLUEBERRY	APRICOT	ASPARAGUS
BRACKEN	BEANS	BEET
CHICORY	BUCKWHEAT	BROCCOLI
CHESTNUT	CHERRY	BRUSSELS SPROUTS
COFFEE	EGGPLANT	CABBAGE
CONIFER	GOOSEBERRY	CARROT
DANDELION	GRAINS	CAULIFLOWER
ENDIVE	GRAPE	CELERY
FENNEL	MUSTARD	CLOVER
FLAX	PARSLEY	CUCUMBER
LUPINE	PARSNIP	LEEKS
MARIGOLD	PEA	LETTUCE
MOSS	PEACH	ONION
OAK	PEAR	SILVERBEET
PECAN	PUMPKIN	SPINACH
POTATO	SOY BEAN	SWISS CHARD
PEANUT	SQUASH	ZUCCHINI
RADISH	STRAWBERRY	
RASPBERRY	TOMATO	
RHUBARB	TURNIP	
SHALLOT		
SWEET POTATO		
TEA		
WATERMELON		

Figure 6.2 Acid and alkaline intolerances of selected crops.

structure are improved by adding large quantities of organic matter.

Micro-organisms in soils

If there is good air–water balance and plenty of organic material, then soils will have trillions of micro-organisms. These are animals ranging from microscopic size to beetles, all eating, breathing, dividing, living, moving, clumping and dying. They aerate the soil, provide water channels, break down large molecules to smaller ones and, in themselves, are a part of the soil nutrient bank and will later provide organic matter for plants. The more organisms the better the soil health and pest management will be. The wider the range of organisms and the larger the population size then the faster the nutrients are cycled and the greater the range of nutrients available to the plants.

Organic matter in soils

It is not really possible to have too much organic matter in a food garden. Organic matter is anything that was once living, and comprises food scraps, grass clippings, hay, straw, leaves, sawdust, and even fur coats, dead cats, jute bags, old cotton curtains and your favourite old jeans. All become part of the soil nutrient bank. As these raw materials are broken down they become humus, a fine, sticky, sweet-smelling, nutrient-rich substance which slowly releases plant and animal micronutrients. Its nature is such that it helps sandy soils to hold water and nutrients and, conversely, helps compacted clay soil to become more open. Organic matter is the very best soil improver. It especially improves soil structure. When soils develop good aeration and balanced water-holding capacity they are said to have a good 'structure'. This looks like open airy bread.

All the soil components are interacting all the time. Soil particles are eroded and dissolved by water. Micro-organisms can't live without water or organic matter. A soil with no organic matter is near death.

Soil abuse

There are several ways in which soil can be abused, all with catastrophic results.

Removal of the surface vegetation

This is the most important cause of decline in soil structure and productivity. To obtain even greater yields and increase the amount of land under cultivation, more and more marginal land has been ploughed up and the vegetation removed. This practice has been disastrous for the soil which, after being dosed with chemicals, is then left exposed and unprotected from animal hoofs, wind, rain, cold and heat. Thousands of years of evolution of soil interaction with plants, animals, air and water is reversed by the removal of surface vegetation and the results are:

- salinity of dryland and irrigation soils
- wind and water soil-eroded land
- toxic soils from chemicals, biocides or nuclear contamination
- soil acidity from overuse of clovers and phosphates
- soil-structure decline from inappropriate farming methods and fertilisers.

Application of artificial fertilisers

If all the fertiliser applied to a crop were taken up by the crop and used for harvestable growth there would be few problems. But, invariably, a significant proportion is not used and is lost. A combination of crop, soil and climatic factors prevents uptake from being complete. For example, rice grown in the tropics uses only 30–40 per cent of fertiliser applied to it. The other 60–70 per cent leaches into ground water, where it is almost impossible to remove, or moves into rivers where it provides the nutrients for various algae, sometimes toxic, which in turn clog the surface and prevent oxygen and sunlight from penetrating. Insoluble phosphates in soils lead to soil acidity (see Figure 6.4).

In addition, excess fertiliser, as mineral salts,

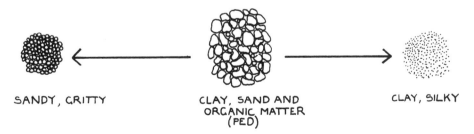

SANDY, GRITTY CLAY, SAND AND ORGANIC MATTER (PED) CLAY, SILKY

Figure 6.3 Soil texture and structure. Sandy soils are made up of large particles and have a rough gritty texture. Clay soils consist of fine particles and feel smooth and silky in texture. In both cases, organic matter will improve the structure and texture of the soil.

destroys the soil micro-organisms and the soil structure, which protects it against wind and water. And where there is loss of vegetation the soil erosion is accelerated. Food plants take up too much fertiliser and cause human sickness, or develop into weak plants susceptible to pest attacks.

Soils with excess fertiliser in them require a 'cleansing crop' which will absorb the surplus fertiliser before food crops are grown in them again. You can grow a cleansing crop of hay, for example, and then revert to food crops. When too much nitrogen fertiliser is used the plants are overly green and the animals and people eating them ingest too much nitrogen. In China where there are relatively high rates of nitrate, nitrite and amines in the diet, correlations are being made with oesophageal cancer (G.R. Conway and J.N. Pretty, *Unwelcome Harvest: Agriculture and Pollution*).

Accumulation of biocides

These are any chemicals used to kill living organisms, and include fungicides, weedicides, miticides and insecticides. They basically wage war against life.

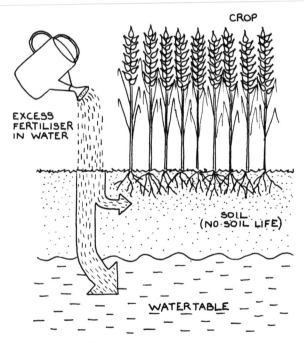

Figure 6.4 *The effect of artificial fertilisers on soil. Excessive use of artificial fertilisers increases the acidity and alkalinity of soil and contaminates underlying water tables. This process is typical of monoculture crop systems which rely on large quantities of artificial fertilisers.*

Many have a very long life in the soil, which means they continue to exist in the soil unchanged because they cannot move into one of the cycles of matter. Or they may move into plants and retain their potency. The shortest life for a biocide is a few hours, whereas others can last up to 40 years.

Bill Mollison recommends that people wishing to grow organic foods use clean farming methods and do not purchase land that has been used for growing bananas, sugar cane, deciduous fruit or orchard crops. A forest of long-term precious timber may work as a cleansing crop. However, by far the worst biocide is nuclear contamination, which persists for years in the soil, plants and water. The site of the 1950s nuclear tests in Australia will have to be locked up for more than 1000 years.

Vegetation clearing and water misuse

Soil salinity is a worldwide problem and occurs because the salts used by the plants for their mineral nutrition, which are normally distributed in correct proportions through the first 2 metres of soil, are concentrated into a narrower layer near the surface. Here the salts accumulate and become toxic. Figure 6.5 shows how the water table, which is normally kept at more than 2 metres from the surface, has risen and concentrated the salts in a much smaller zone as a result of the vegetation being cleared. The crop or pasture then dies. Farmers call it the White Death.

Inappropriate farming methods

Soil structure is destroyed when ploughshares invert the soil and then harrow it to break the clods into a fine tilth. This puts poorer-quality subsoil on top of good topsoil, and allows the fine, now structureless, soil to wash away, blow away and dry out. Today many farmers direct drill seeding into subsoil where last year's stubble remains to protect the topsoil.

In tropical and marginal dry areas, hoofed animals at stocking rates heavier than the earth can support compact the soil to a claypan that becomes impermeable to water. The air spaces in the soil are compressed and the soil becomes like concrete and won't absorb water. This state is called a caleche.

A. BEFORE TREE REMOVAL

CROP GROWING WELL

2m

RECHARGE AREA

WATERTABLE

DISCHARGE AREA

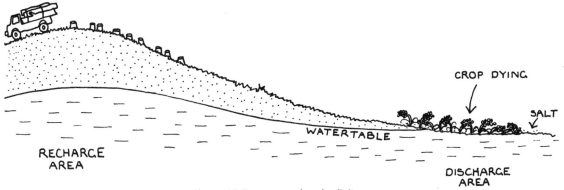

B. AFTER TREE REMOVAL

CROP DYING

SALT

RECHARGE AREA

WATERTABLE

DISCHARGE AREA

Figure 6.5 Tree removal and salinity.

Where heavy ploughing machinery has been used for many years, it always cultivates to the same level and leaves a claypan at the ploughshare depth. This claypan prevents water penetrating and so soils dry out as more water runs off and cannot percolate into the subsoil.

Strategies for repairing and rehabilitating soils

With the right techniques, patience and a lot of hard work, it is possible to rehabilitate abused soils.

Landshaping

Steep sloping land and sandy soils are best protected by terraces that follow the contours of the land. These should be constructed from plants such as vetiver grass, stone or timber risers. The terraces will hold the soil when there is heavy rain, wind or cultivation. Large and small swales function the same way.

Cover soils

Under natural conditions plants shed leaves, which then form a layer of mulch to protect the soil. Grasses and groundcovers grow over them, binding the soil with their adventitious roots as living mulches and protect them. Some shrubs grow branches to the ground and also protect soils. So in permaculture we become conscious of bare soils on farms and in gardens and the many ways to cover and protect them. A pasture should be so covered in grass you cannot see the soil between the clumps or runners. In your garden your spirit should be quiet when all your soils are well covered with appropriate mulches.

Keep water in soils

Soil holds and filters water, cleaning it and giving life to the soil. The first step in soil and land rehabilitation is to get water into the soil. You can use swales and a number of other techniques to slow water down and filter it.

Keep soils in place

Soils naturally move downhill to the bottom of slopes, where they are deeper and richer, leaving thin, dry soils on the tops of slopes. To prevent this, keep soils on the tops of hills covered and manage them so the water moves slowly off the heights and downhill. Also, soils in valleys and along riverbanks are prone to erosion and pollution from fast-moving waters and must be protected by dense plantings.

Altering soil qualities

Generally, alter soils as little as possible, especially chemically. Rather, grow in them what they support. If you have alkaline soils, then grow alkali-loving plants in them and in acid soils, acid-loving plants. Organic matter is the great panacea for all soil problems, whether you want to repair soil or increase its fertility, because it attracts a huge range of micro-organisms, which produce a huge range of nutrients.

Soils can be quite rapidly repaired. In permaculture, soil repair is fundamental to productivity and human health. Different methods are appropriate to different climates, sites and enterprises. You will learn about them when you look at each zone in Part Three. When you design a productive permaculture system, choose the appropriate techniques to repair the soil in gardens, orchards and farms.

How to feed your soil

You can easily achieve high yields without destroying soils or using artificial agricultural fertilisers or biocides. The secret is in supplying large quantities of organic matter. After the first establishment years of your garden or farm, you can become sustainable by growing all the necessary biomass for mulch or compost so you won't have to import or buy any fertiliser or mulch.

Your aim is to use all nutrients completely so there is no surplus, no waste and no pollution. You achieve this by growing a large range of plant species to use all the forms of nutrients, and by applying nutrients at times when they will be most fully utilised.

We will now look at all the ways to supply organic matter and nutrients for soils and plants.

Green manuring

This is when crops are planted specifically in order to be cut and returned to the soil as high-quality organic matter. The green manure crop is slashed two or three times while it is growing and before it flowers and seeds. It is then chopped and incorporated into the soil. This strategy quickly improves the soil texture and structure as well as providing nutrients. In winter you can use species such as rye grass, lupins or barley. In summer you can grow wheat, lucerne (alfalfa) or buckwheat. Use this strategy in orchards and on farms with depleted soils.

Legumes

These are plants that have the bacterium *Rhizobium* living in their roots. The bacteria supply the plant with nitrogen in soluble forms the plant can use, and they excrete surplus nitrogen into the soil around the root zone (the rhizosphere). In return, the bacteria receive energy from the plant. All legumes are 'nitrogen-fixing' if the correct bacterium is present in the soil. You know the bacteria are there if the roots have small white nodules on them and if, when you split the nodules, they are pink inside. The plant and the bacterium are together, symbiotically, nitrogen-fixing. Legumes are vegetables like peas, beans and broad beans, and trees and shrubs like acacias, cassia, leucaena and glyricidia. Many have a pea flower and they all have seedpods that split down both sides. All legumes are soil-improvers and supply nutrients. The leaves of legumes also have 25 per cent more nitrogen in them than other plants. They are usually pioneer plants and prepare the soil for the final species in a succession.

Use the vegetable legumes in a rotation in your food garden and leguminous trees in Zones II and IV.

Cover crops

These are very like green manure crops. They carry out the same function to protect the soil, and give products as well. They are often annuals. Cover crops are not cut because their function is to open up the soil, create a humidity interface, and protect it. Pumpkins and potatoes are good cover crops for hard compacted soils. The root systems of these plants open up the soil for air and water to enter while protecting it from erosion and desiccation. Cover crops are effective in large paddocks but are mostly used in gardens and orchards.

Organic mulches

These cover the soil and moderate summer and winter temperatures by insulating it from extreme heat and cold. They also protect it from erosion by retaining soil moisture, and in addition act as a weed barrier. When mulches are organic materials such as hay, grass clippings, straw, newspaper, old woollen underfelt and so on, they gradually add to the soil organic matter and function as a nutrient bank while they break down.

Living mulches are excellent in big areas and Zone II, while the dead mulches, which require more work, are appropriate for Zone I. They are too much work for the other zones unless you spot-mulch.

Animal manures

Animals are a vital part of the soil nutrient cycle. They carry out many functions, one of which is to supply nutrients in the form of manures. On the whole those animals that eat meat, such as chickens and pigs, have stronger manure (more nitrogen), which requires composting before it is applied to gardens. Cow and horse manure is weaker (less nitrogen) because they are grass-eaters, unless they

have been stabled and have urine (nitrogen) mixed with it. You can place it directly around plants, but it often contains undigested weed seeds.

Poultry grazing in your orchard will, at the right stocking rates, keep it well fertilised. For the other zones (III and IV), use larger animals such as deer, sheep, cows and alpacas, but adjust the stocking rates and rotation so they are effective maintainers of pasture and browsers.

Composts

There are thousands of recipes for compost and every farmer or gardener believes totally in their own method for making it. So here is mine. I like it because it's little effort and it works. It was taught to me by women in Cambodia.

1. Make a 1-cubic-metre box or frame of wire (I have been successful with old iron). Have all your materials ready to fill it.
2. You need a big pile of cream materials like straw or dry grass, another pile of green materials like weeds and grasses, and some manure from chicken or pigs. Chop everything finely.
3. Now make layers. The rule is to use a ratio of 25:15:5. Start with 25 centimetres of finely chopped cream materials (this is high in carbon). Wet the layer with a watering can.
4. Add 15 centimetres of finely chopped green materials. Wet this layer with a watering can.
5. Add 5 centimetres of the manure. Wet it too.
6. Start again and continue until the container is full.

In warm climates you will have compost in 28 days. In cooler climates it takes a bit longer but is still very fast.

I don't turn the compost. When I want to use it, I take off the top few centimetres, which aren't composted, and make them the bottom layer of the next batch. Easy!

Nutrient broths

These are soups for crops. They get a very quick response and are particularly good for plants when you are not sure why they are ailing. There are a number of recipes. Most of them require buckets or barrels of water, or animal manure in a sack, or leaves such are comfrey that are dropped into the barrel and allowed to ferment. When the broth bubbles, the liquid is siphoned off and given to the plants to drink. Broths work by supplying liquids easily absorbed by plants.

Innoculants

Innoculants are a little like nutrient broths. Damaged soils are always low in micro-organisms, which break down large organic soil molecules into those able to be dissolved and absorbed by the plant roots. To make an innoculant, plant and animal matter is fermented to breed up a concentrated supply of micro-organisms. Ferments contain large amounts of fungi and yeasts, which are important in soil nutrition. This is then sprayed on fatigued soils, pasture or crop. The broth supplies the missing organisms and its use must be accompanied by inputs of organic matter, such as manures, mulches and composts, otherwise the organisms will die.

Biofertilisers

These are micro-organisms isolated from the roots of plants and which replenish nutrient salts such as potassium, phosphate and nitrogen in the soil. They are cultured in laboratories and grown on a humic acid substrate. They have enormous potential to transform farming currently dependent on chemical fertilisers. Biofertilisers:

- improve biodiversity in cultivated soils by increasing the populations of naturally occurring groups of soil micro-organisms
- replace 50–100 per cent of inorganic nitrogen (urea)
- increase yields
- improve plant health and reduce pests and diseases
- improve soil fertility.

Nitrate levels decrease in products grown with biofertilisers. High nitrate levels in vegetables are an indirect cause of a modern agricultural disease called methaemoglobinaemia. The causal link has been established. Other links are with bladder, oesophageal and gastric cancer.

Farmers who have become dependent on chemical fertilisers such as urea should at first partially substitute biofertiliser for urea, and then in subsequent years completely replace these chemicals. The arguments for using biofertilisers include:

- field trial results show that biofertiliser together with compost applications gives significantly increased yields—more than biofertiliser alone, and better than chemical NPK applications
- farmers can make biofertiliser themselves and so are self-reliant
- it is very much cheaper than urea
- it encourages good soil conservation and management practices
- it has regenerated exhausted soils
- it is environmentally benign because the organism does not move from the rhizosphere (root zone) while chemical nitrates are often pollutants moving into soil, underground and surface water.

Figure 6.6 summarises the best nutrients for different areas of farms and gardens. This is based on how much work is required, and how easy the materials are to obtain and use.

The future

Chemical fertilisers cost the Earth in fossil fuels and land degradation. All the techniques listed above are effective in restoring soils, and all of them can be implemented without high costs or large machinery. The future for soil rehabilitation lies with soil micro-organisms and large quantities of organic matter, which the micro-organisms require in order to build up their populations. In particular, biofertilisers have huge potential to increase crop yields, improve the texture, structure and water-holding capacity of soils, and provide better soil buffering of acidity and alkalinity than present fertilisers of any type. They also don't pollute soils or water.

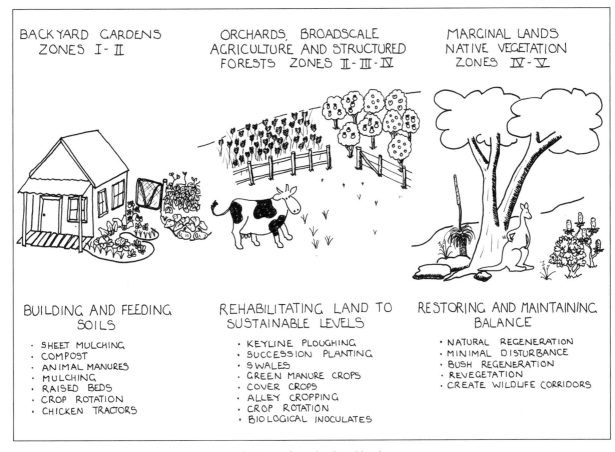

Figure 6.6 Techniques of good soil and land management.

Soil acidity is caused by insoluble phosphate in soils. However, some organisms, known as phosphate solubilising micro-organisms, can solubilise phosphates. By isolating special soil micro-organisms and bacteria, particularly those belonging to genera *Pseudomonas* and *Bacillus*, and fungi belonging to the genera *Penicillium* and *Aspergillus*, which possess the ability to bring insoluble phosphate in soil into soluble form by secreting organic acids, soil acidity can be reversed.

Traditional soil classification system

Before soil scientists were thought of, farmers knew their soils and what they would be capable of, and speedily recognised soil problems. They made field observations which, when put together, gave a comprehensive soil picture. This involved using all the senses to assess soil potential or soil problems and gave rise to the traditional classification system.

As you use the system, you will learn to recognise and remedy many soil problems without complicated analysis. Look at Figure 6.7 and see how many factors interact. When you put these together you will have made a very accurate picture of your soil because factors are interactive in soils and you will be able to consider them simultaneously. This is the best way of recognising soil types and how they work.

Try these:
1. List the plants in your garden and then look at the pH table in the text. Make an educated guess at the soil pH and then write it beside the plant name.
2. Dig up handfuls of soil from three different parts of your garden. On your site analysis plan, mark where they came from and plot the

TRADITIONAL CLASSIFICATIONS OF SOIL

CHARACTERISTIC	INDICATOR OF...
COLOUR	
• COLOURLESS/WHITE:	HIGH SILICA CONTENT
• LIGHT/WHITE:	LACK OF OXYGEN; LEACHED; HIGH CALCIUM CONTENT; ALKALINE pH
• YELLOW:	LACK OF OXYGEN; HIGH CLAY CONTENT; ALUMINIUM AND IRON
• RED:	IRON OXIDE
• RED/BROWN:	VOLCANIC, BASALT ORIGIN; IRON AND MAGNESIUM
• BLACK:	RICH IN ORGANIC MATTER AND NUTRIENTS; HOLDS MOISTURE
VEGETATION	
• EG. AZALEA, BERRIES, CONIFER, DANDELION, DOCK	ACID SOILS; USUALLY LEACHED; OFTEN COMPACTED, WITH POOR DRAINAGE
• EG. SALTBUSH, SPINIFEX, CLOVERS, VETCH	ALKALINE; SALINE; DRY SOILS
• EG. NETTLES	EXCESS NITROGEN; LOW HUMUS CONTENT; LOW MICRO-ORGANISM CONTENT
• EG. BLACK BERRIES	OPEN, DISTURBED SOILS
• EG. BRACKEN, BLADEY GRASS	SOILS RECOVERING FROM FIRE; GENERAL DECLINE IN SOIL FERTILITY
• EG. BUTTERCUP	LOW HUMUS; POOR DRAINAGE
• EG. THISTLES	LOW CALCIUM AND IRON CONTENT; HARD SOILS
PARENT MATERIAL EG. SOILS DERIVED FROM	(AFFECTS STRUCTURE AND TEXTURE)
• SANDSTONE:	SANDY; HIGH SILICA CONTENT
• SHALE:	CLAY; HIGH SILICA AND IRON CONTENT
• BASALT:	HIGH IRON AND MAGNESIUM CONTENT
SMELL	
• SOUR:	LACK OF OXYGEN; SULPHUR DIOXIDE (ROTTEN EGG GAS); ACIDIC
• SWEET AND EARTHY:	HIGH OXYGEN CONTENT; STICKY; CRUMBLY; PROLIFIC SOIL LIFE
• GARLIC:	ARSENIC IN SOIL
TASTE	
• SMOOTH AND SLIPPERY:	ACIDIC; SOIL WATER LATHERS EASILY
• WEAK SODA:	ALKALINE/MINERAL; SOIL WATER WONT LATHER EASILY

SOIL LIFE	
• WORMS :	GOOD MOISTURE CONTENT ; RICH IN ORGANIC MATTER AND NUTRIENTS ; LOW PESTICIDE CONTENT
• ANTS :	DRIER, SANDY SOILS WITH LOOSE TEXTURE
• SLUGS AND SNAILS :	DAMP AREAS ; OPEN LOOSE MULCHES
• SKINKS AND LIZARDS :	DIVERSITY OF RESIDENT GARDEN INSECT LIFE
• WOMBATS :	DEEP, SOFT, MOIST SOILS

HOW SOIL HANDLES WATER	
• RUN - OFF :	BARE GROUND; COMPACTED SOILS ; SLOPE TOO SEVERE
• WATER REPELLENT :	COMPACTED ; ERODED ; EXCESSIVE USE OF DOLOMITE
• SHRINKS AND SWELLS :	HIGH CLAY CONTENT ; HOLDS WATER ; NUTRIENTS OFTEN TIGHTLY BOUND ; NOT GOOD FOR HOUSE FOUNDATIONS ; CAN CRACK BUILDINGS ; MAY BE USEFUL FOR MUDBRICKS
• FAST DRAINING : IE. A HOLE FILLED WITH WATER THAT DRAINS WITHIN TEN MINUTES IS CONSIDERED TOO FAST FOR GOOD PLANT GROWTH	EASILY ERODED ; COLLAPSES EASILY ; FEW FUNGAL DISEASES ; MICRO - ORGANISMS AND NUTRIENTS MOVE QUICKLY UP AND DOWN SOIL PROFILE ; NOT GOOD FOR DAMS

HISTORY	
• BARE GROUND :	POSSIBLE AGRICULTURAL OR INDUSTRIAL CHEMICAL CONTAMINATION
• GROWTH IN POOR SOILS :	POSSIBLE SITE OF OLD CHICKEN PENS, PIG YARDS OR HORSE STABLES
• NO TOPSOIL :	SITE COULD HAVE BEEN USED AS A QUARRY OR FOR FILL
• BAD CRACKS AND RUBBISH :	POSSIBLE SITE OF OLD TIP OR LANDFILL

different soils. Write down the texture of each one. Look at how the particles stick together then wet them and see if they roll into 'worms'. This is a test of soil structure.

3. How would you make one of your soils into a highly productive garden soil?

4. Bury a bucket of kitchen waste in your garden. After three weeks dig it up and see how many animal residents you can identify there.

5. Carry out a whole site soil analysis. Draw a 'mud' map and use coloured pencils to show the different soils in the different microclimates. Think about how you will repair them if they are damaged.

CHAPTER 7

Trees, forests and windbreaks

A forest cannot be precisely measured or costed. Neither can its destruction. However, we do know that removing forests results in water loss, nutrient loss, soil loss, salinity problems, river flooding, local drought, habitat loss and destabilising climate. How do we cost these?

Forests behave differently at different ages. Old forests are anchored and strong and give back 25 per cent of themselves annually as nutrient in leaves, roots, seeds and flowers. They are generous and prolific. Old forests have special functions in relation to wind, rainfall, dust, nutrient turnover and protection. Young forests are different. A young forest is a voracious consumer of nutrients and gives little back to soils and other plants, and although it traps large quantities of carbon dioxide, this does not even begin to compare with the amount of carbon tied up in an old forest. Clear-felling forests takes all the big old trees and those left are particularly susceptible to diseases and climatic catastrophes such as huge storms, hail and drought.

Saving old forests and planting new ones is still the best way to tie up the surplus atmospheric carbon that causes global warming, and to reduce the heat reflected back from bare earth and other surfaces into the atmosphere. Scientists are now saying that when the world's forest area drops below 40 per cent of its original cover, then other systems will break down. This 40 per cent seems to be a critical point for sustainability of water, soil and rivers.

 Our ethical task is to:

- save all the remaining old forests
- plant more trees for increased condensation on rainy, windward slopes
- plant forests to maintain water purity in rivers, lakes, ponds, dams and streams
- replace all forests for the products we have used
- plant forests for future generations
- plant forests to stabilise natural and cultivated ecosystems.

 Our forestry and windbreak design aims are to:

- design and plant perennial, diverse vegetated landscapes so they work with the wonder, efficiency and diversity of a guild, or *waru*
- design landscape sites to achieve at least 40 per cent tree cover at maturity.

 If we don't have design aims for forestry and windbreaks:

- fresh water is polluted
- rivers and lakes silt up and flood

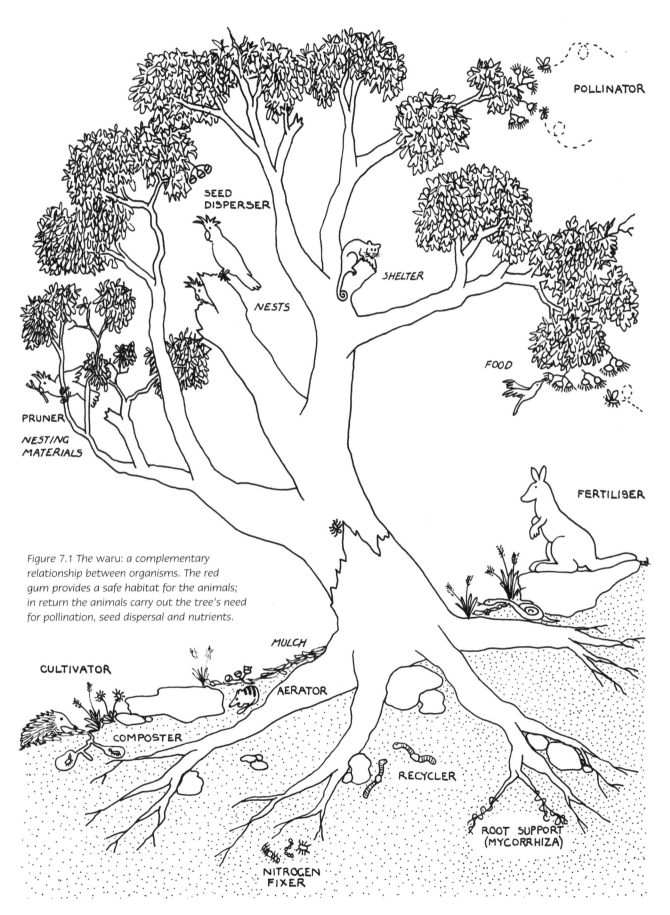

POLLINATOR

SEED
DISPERSER

SHELTER

NESTS

FOOD

PRUNER

NESTING
MATERIALS

FERTILISER

Figure 7.1 The waru: a complementary
relationship between organisms. The red
gum provides a safe habitat for the animals;
in return the animals carry out the tree's need
for pollination, seed dispersal and nutrients.

MULCH

CULTIVATOR

AERATOR

COMPOSTER

RECYCLER

ROOT SUPPORT
(MYCORRHIZA)

NITROGEN
FIXER

- soil water is depleted
- animal and crop productivity decreases
- energy costs are higher
- water leaves land fast, and soil erodes easily
- windborne diseases and weed seeds are not filtered by vegetation.

The functions of a forest

You can think of forests as living organisms carrying out special and unique processes for life to continue on Earth. The primary purpose of forests is to give soils the time and the means to hold and cleanse water on land before it moves to rivers, lakes and aquifers. It requires an entire forest to do this, with each individual tree carrying out its co-operative function to achieve the purpose of the forest.

The forest as a co-operative

Like all organisms, a forest has several parts. Trees, shrubs, herbs, grasses and groundcovers are the *fixed* species of a forest and they carry out special functions. To do this they need pollinating, seed dispersal, fertilising and cultivating. Other species are *mobile*, like birds, insects, mammals and spiders. The mobile species work for the fixed species and the fixed species provide them with food, safety, medicine and nesting materials. Together the tree and its associates can be imagined as a guild, or *waru* (see Figure 7.1).

A large planting area of monospecific trees such as pines is only a plantation and not a forest because of its limited functions, habitats and components.

The cells of the forest

When you visualise an old forest as a single organism, then each tree can be seen as cells of the forest, each one shedding its own weight many times in its lifetime. Once we understand the way they work, we can see that trees are miracles of engineering.

Trees are far from uniform and each tree's many parts are like mini-ecosystems. The bark, the roots, the flowers, leaves and growing points comprise quite different zones of one tree, and perform quite different functions. To understand the forest we need to understand how trees work with wind, sun and water—the components of climate that you read about in Chapter 5. Trees are superb regulators of air quality, temperature, humidity and wind.

Forests and wind

- *Wind pruning and anchorage:* Trees are 'pruned' or deformed by prevailing winds and from this you can fairly accurately predict local wind direction, intensity and places for windbreaks. Heavy trees with large canopies such as oaks rely mainly on their weight to withstand severe winds. Other trees with lighter canopies insert roots deep into the ground and anchor themselves. It is important to use anchoring trees in cyclone areas.
- *Trees on the edge of forests:* Trees with light-coloured bark and leaves grow on the prevailing wind side and this light colour deflects both wind and light to some extent. In deciduous forest this role is performed by the birch, and in Australia, light-barked eucalypts.
- *The edge as a wind filter:* Wind, like water, carries a 'load'. It carries ice particles, sand, dust, bacteria, viruses and seed. Some trees, especially on the windward side, have a thickened bark to withstand particle blast. Because trees with small fine leaves can 'capture' the load and deposit it as nutrients, the edge of a forest can be seen as a nutrient trap, and the edge facing the prevailing wind will have richer soils and hardier plants than the edge on the leeward side of a forest. Figure 7.2 shows how the edge must be kept permanently intact because if it is destroyed then windburn, abrasion, disease, pests and weeds enter the forest and destroy its integrity.

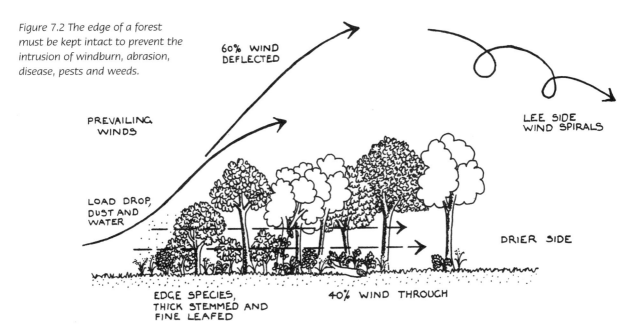

Figure 7.2 The edge of a forest must be kept intact to prevent the intrusion of windburn, abrasion, disease, pests and weeds.

- *Deflecting winds:* Typically, about 60 per cent of the wind stream is deflected up and over the forest. This is because of the specialised forest edge which is essential to the lift of the wind. Edge species are usually pruned to a 45-degree or 60-degree angle, and these species are dense, small-leafed and tough with thick stems. The 40 per cent of wind that penetrates the 'edge', or forest closure, is cleaned of most of its load and its energy is absorbed. The lifted and deflected wind is then compressed in a belt up to 20 times the height of the tree canopy. If this is humid air then it is compressed again, cools, condenses and it rains.

- *Modifying wind temperature and humidity:* If the wind is cold, it is warmed by condensation as it passes over the leaves of trees and shrubs. If the wind is hot, it is cooled by evaporation. Within 500 metres of the edge the wind comes to stillness. At this point in the forest the air is clean, warm, still and slightly humidified. This is a perfect growing place. It is what we want to design for intensive, productive, protected growing and, like this place, we need to farm in forest clearings.

Trees and sun

- *Photosynthesis:* Trees absorb the sun's light energy and turn it into chemical energy by the process called photosynthesis. Where leaves are dark green or reddish, as often found in the tropics, more light is absorbed and local temperatures are reduced. Photosynthesis requires water from the soil. Evaporation and transpiration (sweating) from the underside of leaves works as a pump to pull up the water.

- *Evaporation and transpiration:* Trees evaporate and transpire water into the atmosphere as humidity. This evaporation is accompanied by cooling so that by day it is cooler in and near a forest than it is in unvegetated areas. At night in humid conditions, moisture condenses on the leaves and warms the surrounding air. Because leaves are 86 per cent water, they are cooler by day and warmer by night than bare ground (the microclimate effect of a water body). Figure 7.3 shows evaporation and transpiration from plants. Together they are called evapotranspiration. In very dry areas, the evapotranspiration from trees humidifies dry air, and in very damp areas water captured by leaves dries the air.

Figure 7.3 Photosynthesis and water movement in plants.

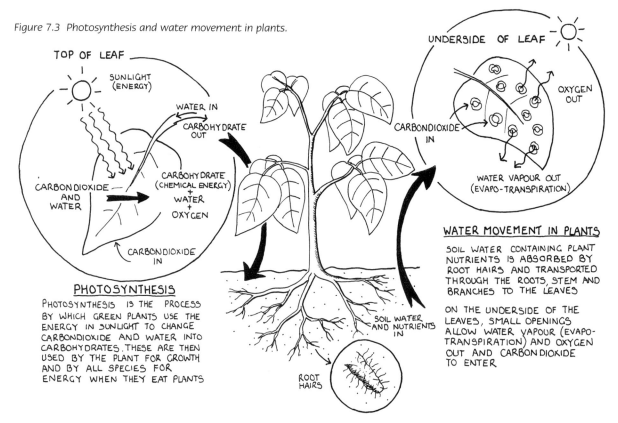

TOP OF LEAF

SUNLIGHT (ENERGY)

WATER IN

CARBOHYDRATE OUT

CARBON DIOXIDE AND WATER

CARBOHYDRATE (CHEMICAL ENERGY) + WATER + OXYGEN

CARBON DIOXIDE IN

PHOTOSYNTHESIS

PHOTOSYNTHESIS IS THE PROCESS BY WHICH GREEN PLANTS USE THE ENERGY IN SUNLIGHT TO CHANGE CARBON DIOXIDE AND WATER INTO CARBOHYDRATES. THESE ARE THEN USED BY THE PLANT FOR GROWTH AND BY ALL SPECIES FOR ENERGY WHEN THEY EAT PLANTS

ROOT HAIRS

SOIL WATER AND NUTRIENTS IN

UNDERSIDE OF LEAF

OXYGEN OUT

CARBON DIOXIDE IN

WATER VAPOUR OUT (EVAPO-TRANSPIRATION)

WATER MOVEMENT IN PLANTS

SOIL WATER CONTAINING PLANT NUTRIENTS IS ABSORBED BY ROOT HAIRS AND TRANSPORTED THROUGH THE ROOTS, STEM AND BRANCHES TO THE LEAVES

ON THE UNDERSIDE OF THE LEAVES, SMALL OPENINGS ALLOW WATER VAPOUR (EVAPO-TRANSPIRATION) AND OXYGEN OUT AND CARBON DIOXIDE TO ENTER

Working in this way, forests can be seen as natural airconditioners that cleanse the air and regulate extremes of humidity and temperature. And, as you know from Chapter 6, tree roots pull up water and keep the water table low, so preventing the salts, which cause soil salinity, from concentrating in the top few centimetres of soil. By absorbing sunlight trees also reduce glare.

Trees and precipitation

- *Condensation drip forests:* Where the airstream is very humid (for example, on sea-facing coasts and islands), the air flows rapidly and condenses on leaf surfaces. Dense rainforests grow from the condensation harvested from leaf surfaces, which can be 80–86 per cent of the total precipitation. When you consider that a single tree can present 16 hectares of leaf surface, a forest has huge potential to capture water through condensation even if it doesn't rain.

- *Trees as evaporative pumps:* Trees pump moisture into the air as they transpire and return 75 per cent of received precipitation this way. The Tasmanian blue gum, *Eucalyptus globulus*, pumps 4000 litres into the atmosphere per day and averages about 60 trees to a hectare in a mixed forest. This is a huge return of water to an airstream, which can condense and rain elsewhere. It has been calculated that as much as 60 per cent of inland water comes from forest transpiration. Therefore, forest removal in one area can relate directly to drought in another.

- *Trees protect soils:* The forest canopy protects soils from water and wind erosion. Bare earth can lose 80 tonnes per hectare of soil in one heavy deluge. Topsoil and organic matter is removed first. This often ends up in the sea and is irretrievable. In addition, the topsoil and subsoil dry out and become hard like a claypan. And with unimpeded run-off, rivers flood and dams silt up.

Rain over forests

- *The interception layer:* When it rains over a forest the rain is spread, as a film of water bound by surface tension, over all the leaves of the trees and is caught in stems, bark, cobwebs, flowers, and insect and bird nests. Some evaporates from these places. The amount caught in the canopy is influenced by the crown's thickness, density of tree canopy, branches and trunk. For 100 per cent of rain falling, 10–25 per cent is caught in the canopy, called the interception layer, without ever reaching the ground. More is caught in evergreen trees than deciduous (see Figure 7.4).

● *The throughfall:* The rest of the rain drifts through the canopy as mist and droplets. It contains organic salts, dust, plant exudates, insect droppings and sheddings. It is a nutrient-rich soup and is directed towards the outer leaf canopy, also known as the dripline, below which are the feeding roots.

● *Mulch blotter:* Before the water can reach the roots, however, bark, taproots, fungi and the humus layer of the soil act like a great water blotter, soaking up l centimetre of rain for every 3 centimetres of depth and holding it for release or use when the soil begins to dry out again.

● *Filtering through soil:* Through the next 40–60 centimetres of soil the throughfall is absorbed into water and air channels, nests and burrows; absorbed by more soil fungi and bacteria; filtered by humus and mineral particles; and, of course, taken up by the tree roots. This water is first bound by particles of clay and humus and then the excess percolates slowly through the soil. At any time some of this water is available to soil organisms. Some water is bound and held firmly, some is stored in cavities in the soil and in humus.

● *Filtered water:* Once all this has been accomplished, excess water starts very slowly to move to rivers, springs and the sea. When it does, it is clean.

Windbreaks

You draw on your knowledge of how forests function when you design windbreaks. Well-designed windbreaks protect land and increase crop yields by carrying out similar functions to the forest edge. They also provide additional yields such as bee fodder, firewood and building timbers. A line of pine trees is not an efficient windbreak because once the lower branches fall off, the wind velocity under the trees is increased and the long black tree shadows reduce productivity.

Wind is fluid and, like water, it can be deflected sideways and lifted. It forms layers naturally

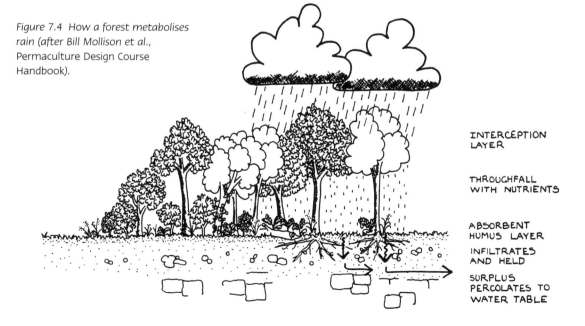

Figure 7.4 How a forest metabolises rain (after Bill Mollison et al., Permaculture Design Course Handbook).

INTERCEPTION LAYER

THROUGHFALL WITH NUTRIENTS

ABSORBENT HUMUS LAYER

INFILTRATES AND HELD

SURPLUS PERCOLATES TO WATER TABLE

because hot air rises and cooler air flows underneath it. Every site has a predictable wind pattern. Sometimes you can learn about it from weather records. In other cases you will need to observe how tree shapes are deformed (wind-pruned) and the amount of wear on buildings.

Each windbreak design is site-specific. You decide where windbreaks are needed from your sector analysis and from your microclimate analysis. Your task in this chapter is to identify where the harshest winds—hot or cold or strong—come from on your land, how long they last and in what season they arrive.

- Design windbreaks as part of the 35 per cent of permanent tree cover needed for each site.
- Use the natural characteristics of winds to create desired microclimates.

When you don't have windbreaks

Cold winds remove heat from the surface of plants, buildings, water and living bodies. With the increase in wind speed and the evaporation of fluids, a chill factor is created. The result is that on windy sites the climate will actually be cooler than the temperature figures show. Plant growth is retarded and both height and yield decrease, while solar devices and insulation work less efficiently on buildings.

The ecological functions of windbreaks are to:

- serve as suntraps
- increase wind velocity or cooling
- decrease evaporation from water or land
- control erosion
- provide shelterbelts for stock
- act as dust filters
- form nutrient traps for wind and water.

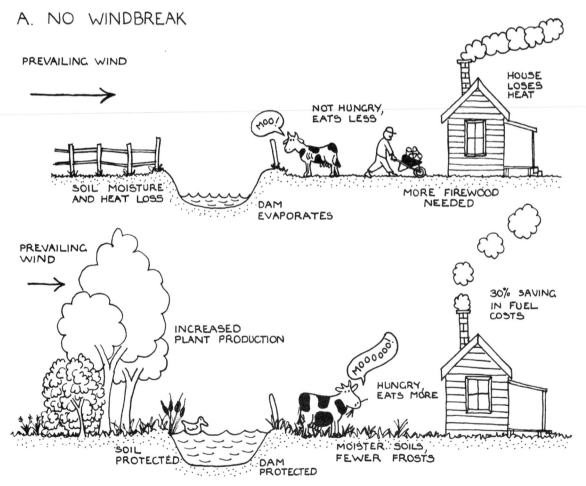

Figure 7.5 Advantages of windbreaks. Climate modification, improved plant and animal production, and energy conservation are some of the benefits provided by windbreaks.

TABLE 7.1: SPECIFIC ADVANTAGES OF WINDBREAKS

Advantages	Examples and design techniques
Protected animals	Windbreaks protect animals from harsh winds during hot and cold weather. In very cold weather animals eat 16 per cent less and in summer heat also have a reduced intake. Ideally, no grazing animal should be more than 400 metres from good shade or the losses in heat and moisture are greater than gains in meat and wool production. (Animals require shelterbelts which are more like open woodlands than dense forest and have a canopy closure of about 70 per cent.)
Protected soils	These retain more moisture and frosts are reduced. Soil on slopes is held better and not as susceptible to wind and water erosion. Soil temperatures are lower in summer and warmer in winter and soils lose less moisture.
Reduced plant damage	With windbreaks, damage in citrus orchards is reduced by 50 per cent. In all orchards there is increased blossom and fruit set, increased pollination and less breaking of branches and uprooting. The overall increase in production is about 25 per cent. The most sensitive plants to wind damage are citrus, avocado, kiwifruit, deciduous fruit, corn, sugar cane and bananas.
Reduced energy loss from buildings	Houses can lose up to 60 per cent of their warmth in winter. Savings of 30 per cent in heating fuels are usual with windbreaks, even in moderate climates. In hot climates large evergreen shady trees designed in an avenue can bring cooled air into the home. Shady trees over roofs and on western walls are also effective.
Cooling very hot sites	Specially designed windbreaks can speed, channel and cool air that is uncomfortably hot for plants and animals.
Protect human settlements	Windbreaks prevent snowdrift in cold climates. They protect roads from high-velocity winds and have an impact on human health. In dry climates they filter the dust that causes ear, eye, nose and lung sicknesses in people. There is some evidence that human and animal epidemics, known to be carried by winds, spread faster when there are no encircling forests or windbreaks around human settlements.
Windbreaks for energy generation	Designed windbreaks can increase the wind speed by directing it to the generator.

Height, density and shape of windbreaks

Figure 7.6 shows how wind moves when it is fully blocked and how it can be directed up and away from soft areas or plantings. A windbreak is effective to a distance along the ground equivalent to about 20 times its maximum height. However, its effectiveness decreases the further away you move from the actual windbreak. There must be some movement of air through a windbreak or the wind eddies, often quite destructively, on the other side of the barrier. The principle is to create the equivalent of the forest 'edge', which will lift the wind. The windbreak can be shaped so it tapers at the ends, reducing wind velocity.

The most effective shape for a windbreak is a boomerang or parabola. This directs the wind off to the sides and also functions as a suntrap. A site can have several of these. The wind, once lifted on the prevailing-wind side, is kept high. This is particularly effective over orchards.

How to design your windbreak

Start with successional planting, as discussed in Chapter 3. Plant small tough species, including local nurse and pioneer species, which prepare the soil for the climax species. Plants with the following characteristics should be included in your windbreak design:

- hardy with deep anchoring root systems
- fire- and wind-resistant with fibrous stems and fleshy or small hairy leaves
- plants with fast early growth—pioneers will fulfil this role
- nitrogen-fixing plants with good leaf fall, which are self-mulching and have additional yields.

Assess the risks to your new plants and, if you require it, provide extra protection with tree guards or fence the whole area.

Windbreaks for different ecosystems and climates

Coastal areas, desert areas, inland regions of heavy frost, subtropical regions in danger of cyclones, and cool climates lend themselves to special windbreak designs (see Figure 7.7). From your analysis you know which way the winds blow in different seasons—which are mild and warm; which are cold and harsh. Perhaps you need dust-filtering windbreaks on the sunny aspect and they may need to be deciduous to receive winter sun.

Small windbreaks

Although you may not need to design and plant windbreaks for large areas, windbreaks are necessary in most gardens, although they need not be permanent. For example, Jerusalem artichokes make an excellent summer windbreak or suntrap in cool climates. They can collect and direct sun to ripen tomatoes.

Windbreaks can be small hedges of herbs only knee-high in some cases. For example, lavender could be used for this purpose in a cool climate. Near the coast where wind is abrasive, desiccating and saline, small windbreaks can be sufficient to raise a summer or winter crop inside them.

A. BRICK WALL

B. ALL EVERGREENS

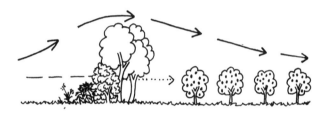

C. PENETRABLE WINDBREAK

Figure 7.6 *Penetrability of windbreaks. A solid barrier increases air turbulence on the leeward side, whereas a porous windbreak, consisting of mixed plantings, reduces wind speed and velocity.*

 Try these:

1. Twice a day, in the early morning and in the late afternoon at sunset, walk around your garden or land and feel with your face or hands where the local air currents are. Sketch these on your plan.
2. Design a windbreak for a site you are very familiar with.
3. Think about whether you need permanent windbreaks or just at certain seasons. Make up a short list of possible plant species.

4. If you were the local town planner, where would you want windbreaks:
   - for shopping in comfort
   - to rest from shopping
   - to eat a picnic with children
   - to walk to church, the post office or other institution?

   What is the negative effect that you wish to moderate—hot dry winds, cold harsh winds, or dusty winds? What species and plant characteristics would you need?
5. What do you think is happening in the city when people talk about the 'canyon' effects of winds?

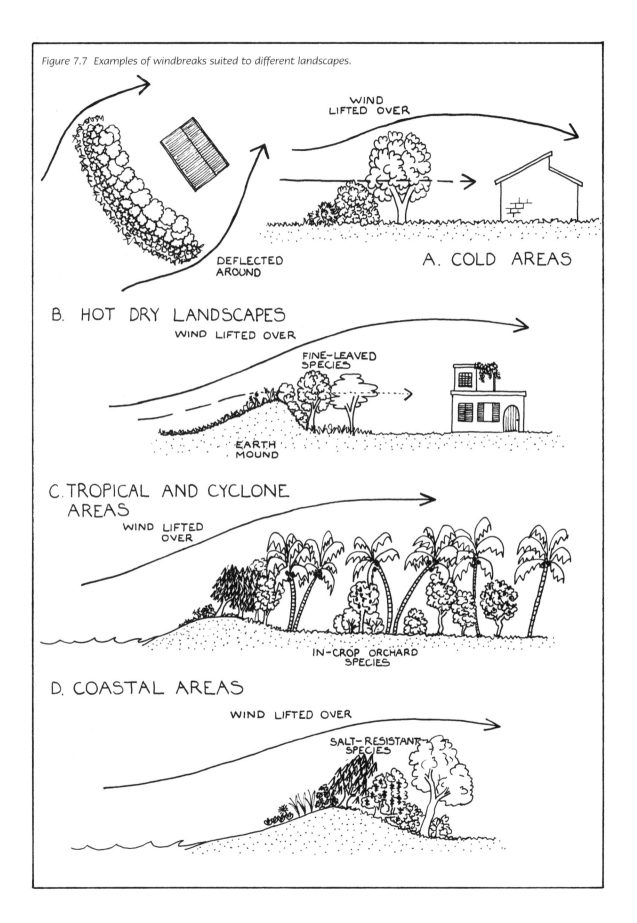

Figure 7.7 Examples of windbreaks suited to different landscapes.

WIND
LIFTED OVER

DEFLECTED
AROUND

A. COLD AREAS

B. HOT DRY LANDSCAPES

WIND LIFTED OVER

FINE-LEAVED
SPECIES

EARTH
MOUND

C. TROPICAL AND CYCLONE
AREAS

WIND LIFTED
OVER

IN-CROP ORCHARD
SPECIES

D. COASTAL AREAS

WIND LIFTED OVER

SALT-RESISTANT
SPECIES

CHAPTER 8

Our plant and seed heritage

Plants form one kingdom of living things. This kingdom is divided up into phyla, families, genera, species and cultivars. A species is generally defined as being able to cross-pollinate and produce viable young, and has two names; for example *Vica faba*, which is the broad bean—*Vica* is the genus name, like a family name, and *faba* is the species name. A cultivar has another name added to the species name; for example, *Vica faba* 'blue gem'.

This system of plant classification is based on similarities of flowers and fruit, and was developed by Carl von Linne in Sweden in 1753. So a tall tree living in the tropics can be closely related to a groundcover in a cool temperate area. For example, *Grevillea robusta*, a large tropical timber tree, is closely related to *Grevillea laurifolia*, a groundcover in cool mountains.

Every country is rich in plants exquisitely adapted to its soils, climate and landforms. This rich and precious natural vegetation cares for soil, water and animal species in ways that we barely understand. Plants, their yields and uses, are a fundamental theme that runs through all permaculture practices. In many places in the world, both indigenous and introduced plants are threatened with local extinction. These plants are morally the public property of all the people who live in that bioregion.

Today our plant heritage is under threat like

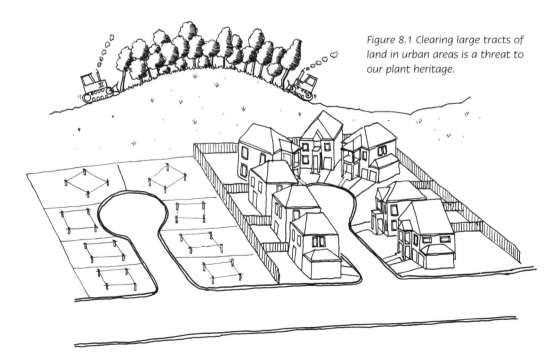

Figure 8.1 Clearing large tracts of land in urban areas is a threat to our plant heritage.

never before. Multinationals produce hybrid species of initially high-yielding varieties (HYVs) that require accompanying chemical fertilisers and biocides. These HYVs require the purchase of fresh seed every growing season because hybrids don't produce seed of the same high quality as their parents. In the meantime, proven and sturdy local varieties are abandoned. The result is that 250 million years of genetic richness is rapidly being reduced.

Clearing land with slashers, chainsaws and tractors for monoculture, housing and lawns destroys plant species indiscriminately. Similarly, cutting down forests and trees for their timber use (especially woodchips) has a profound impact on species diversity and the health of our ecosystems. Plants are also under threat in smaller ways; for instance, you can mow a plant to death.

Our ethical task is to:

- preserve and propagate all local and traditional introduced species, and to save and grow their seed
- start local seedbanks and contribute to existing ones.

Our design aims for plants and seeds are to:

- know where to place them in a permaculture design
- deduce how climate and microclimate will affect them
- describe their functions and products
- grow non-hybrid species by simple propagation techniques
- select, dry, store and grow our own seed.

If we don't have design aims for plants:

- local species are lost, some before they are even identified
- plants are placed wrongly and the design is weakened
- disasters destroy all our crops—for example, flood, fire or ruthless slashing

- we have outbreaks of pests and diseases
- international companies can steal our plants, seeds, genes and chromosomes
- we lose our choice of species and cultivars and local species.

The function of plants

As permaculture designers we are interested in how plants function in a design, and what their yields and characteristics are (see Figure 8.2). We also classify plants by these functions, yields and characteristics. Citrus trees are a good example of this; they are found in several plant lists according to design needs. Under *characteristics* we could list that they are evergreen, leafed to the ground, bear fruit, and have aromatic leaves and flowers. Their *functions* would include their ability to be used as a windbreak for the cold side of a site or as part of a suntrap, and the fact that they're self-mulching. Citrus trees also *yield* whole fruit, jam, juice, peel, seeds, aromatic oils and a timber highly prized for altar tryptichs.

By knowing the characteristics, functions and yields of various plants, you are then able to choose the best plants for your specific site. For instance, to plant a windbreak you would list all the plants that function as windbreaks and have the following desirable characteristics:

- leafed to the ground
- multi-stemmed
- bun-shaped
- small leaves
- evergreen.

Once you recognise a specific site need—such as drought-resistance or acidity-tolerance—then the question is, 'What is my choice of plants?' Every plant must carry out two or more functions. There are websites with extensive lists for how plants function in permaculture systems (see References).

Plant diversity

Earlier, in Chapter 3, we saw how species diversity gives stability to ecosystems. It is particularly important in plants because they have a wide range

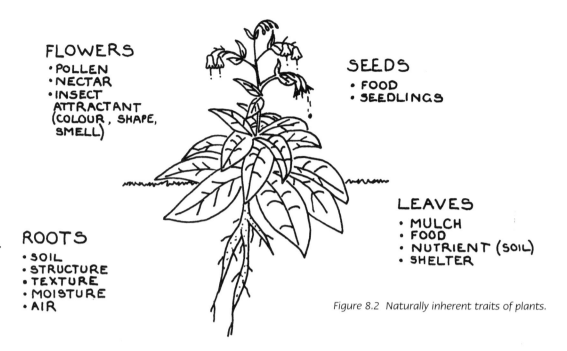

FLOWERS
- POLLEN
- NECTAR
- INSECT ATTRACTANT (COLOUR, SHAPE, SMELL)

SEEDS
- FOOD
- SEEDLINGS

ROOTS
- SOIL
- STRUCTURE
- TEXTURE
- MOISTURE
- AIR

LEAVES
- MULCH
- FOOD
- NUTRIENT (SOIL)
- SHELTER

Figure 8.2 Naturally inherent traits of plants.

of useful and wonderful characteristics and functions, including:

- pest resistance
- nitrogen supply
- nutrient cycling
- fire resistance
- efficient water usage
- carbon sink
- glare reduction
- mulch
- temperature modification.
- drought survival
- soil improvement
- windbreak
- variety of yields
- soil protection
- shade
- flood resistance
- habitat

In the event of major or minor disasters—from floods and droughts to a partner going berserk with a slasher—if you plant a wide diversity some plants will survive. Aim for diversity of cultivars (varieties) as well as diversity of species. For example, if you plant four or five varieties of onions you will achieve continuous yields and also ensure that at least some will survive a disease epidemic or other environmental disaster.

Identifying plants

One problem for designers is identifying and obtaining varieties of plants that grow well in the local area. These plants are not usually available from commercial outlets but older gardeners will tell you about them and very often give you seeds,

cuttings or seedlings. Local herb and garden clubs hold a lot of knowledge as well.

Plants can be grouped according to their climate of origin. Some fruits and vegetables thrive in oasis climates and Mediterranean climates with hot dry summers and cold wet winters. But these plants tend to get fungal diseases when grown in places with warm and wet summers. However, you can have success if you plant them in special microclimates, such as next to a western wall with some air circulation.

When you find a new plant you want to identify for use in your permaculture system, look at it very carefully. Plant groups tend to have distinctive characteristics; for example, all the thyme family have small hairy leaves. You can probably see that an unknown plant is like some other plant you know. Think about it and then check the features around it. Notice the soil, the aspect, the slope. Is it a tree, shrub, herb or grass? Identify its yields and functions—mulch, groundcover, shelter, food and so on. Use your natural senses.

- *Touch it*—feel the texture of the leaves and examine the flower or fruit if it has any.
- *Smell it*—crush the leaves and see if the scent reminds you of other plants you know. All the mints are identifiable by smell, and so are the lavenders, eucalypts and citrus.

- *Taste it*—chew the leaf and spit it out, and again see what it reminds you of; for example, the oxalis family all have the same acid taste, as do the sorrel. (It is very hard to poison yourself from simply doing a test taste and then spitting out.)

Major food plant groups

Plant groups have many family members which like the same growing conditions. So, if you know one member of a family you can guess where another member would like to grow.

In general, each plant family member requires similar conditions. For example, cabbages grow better in winter and they are gross feeders, which means that they need very rich soils. Tomatoes are mainly summer crops (except in the hot, wet tropics, where they are winter crops) and need a fairly long growing season. When you know this you can plant your food supply and plan its continuity and what each plant needs.

Two important families

- *The legume family:* Members of this family live in symbiosis with nitrogen-fixing bacteria on their roots. They provide soluble nitrogen to the plant and give surplus to the rhizosphere (areas around the roots) and to the soil. They can be thought of as nutrient-supplying plants. They also supplement

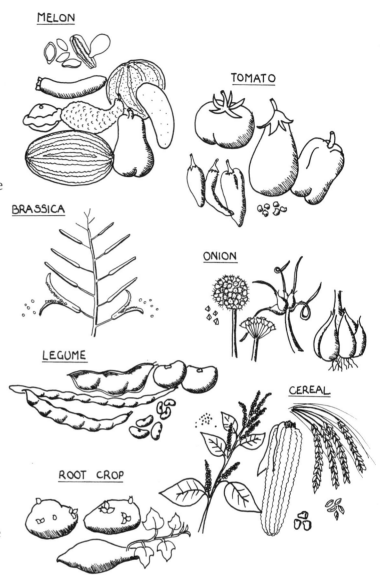

Figure 8.3 Fruits of the major food plant groups.

TABLE 8.1: MAJOR FOOD PLANT GROUPS

Family	Some family members
Melon	watermelon, rockmelon, honeydew, cantaloupe, cucumber, squash, pumpkins, zucchini and gourds
Tomato	tomatoes, eggplant (aubergine), chillies, sweet peppers (capsicum), pepino
Cabbage (also known as Brassica)	cabbages, broccoli, Brussel sprouts, collards, rape, mustard, cauliflower and kale
Onion	onions, leeks, garlic, chives, garlic chives, shallots and spring onions
Legume	peas, beans, lupins, lentils, chickpeas, mung beans and soybeans
Cereal	sweet corn, flour corn, wheat, rye, buckwheat, barley, rice, oats, sorghum
Root crop	turnips, swedes, carrots, parsnips, radishes and potatoes

cereals or grains to provide protein for diets. Legumes are an important part of the diet of vegetarians and vegans, and people in developing countries.

- *The cereal family:* These are also known as staples because they form a very large part of all people's diets and provide them with energy or carbohydrates. There are local staples in all countries and cultures. For example, Scots eat rye and barley, Chinese eat rice, Irish eat potatoes, and South Africans eat corn.

Propagation

Growing new young plants from old ones is called propagation. When left alone in their natural environment most plants reproduce themselves perfectly well, even adapting and growing more strongly. However, some cultivated plants require special techniques to reproduce. Plants reproduce themselves in several ways; the main ones are listed in Table 8.2. You can read about these methods in

any good book on gardening. They are fun to try and you can give your surplus plants away as gifts.

Seed-saving

Seed-saving is cheap, easy to learn and to practise, and needs only a little equipment. Seed-saving means collecting your own seeds to:

- grow the seed well
- protect the seed from going bad
- keep it for a long time, if desired
- save money usually spent on buying seed
- have your own choice of varieties
- save the traditional, heritage varieties
- keep seed that is good for different conditions, such as drought, flood, disease
- breed new varieties
- share seed or swap with neighbours
- select for local qualities, such as high yielding, low compact plants
- have very good-quality seed
- have seed at home for next season's planting.

People who save seed don't need to buy it in the

TABLE 8.2: PROPAGATION METHODS

Method	Techniques
Seed	When plants are grown from seed, half the genes of each parent combine to make one new individual, so each new plant is slightly different from its parents.
	Most annuals and some biennials are grown from seed. All the vegetables except potatoes—which grow from 'seed' potatoes, kept for that purpose—grow from seeds.
Cuttings	Cuttings are taken from trees and shrubs. As this is a type of cloning, each new plant will be almost identical to its parent. *Soft-tip cuttings:* Small shrubs like herbs—lavender and rosemary—are grown from soft-tip cuttings. These are the young fast-growing tips cut after the plant has finished flowering in spring or summer. *Hardwood cuttings:* Pieces about 30 centimetres long are taken in winter when about as thick as your thumb. These are planted in a pot or the ground and kept wet until they shoot in summer. Many deciduous plants, like roses, figs, grapes and mulberries, are grown from hardwood cuttings.
Budding and grafting	Budding involves taking a bud and splicing it into the bark of a tree with a strong root system. Grafting involves taking a small branch and splicing it onto a tree with a strong root system. There are many methods of grafting. Both budding and grafting are usual where there is a very good fruit without a strong root system. All citrus in Australia are budded onto rough lemon rootstock. Deciduous fruits are generally grafted.
Division	This method is common for plants with underground stems or tubers. With potatoes, which are underground stems, pieces are taken with an 'eye', or growing point, and simply planted. With chives, parent plants produce bulbs, or 'off-sets', and in spring these can be divided and planted out.

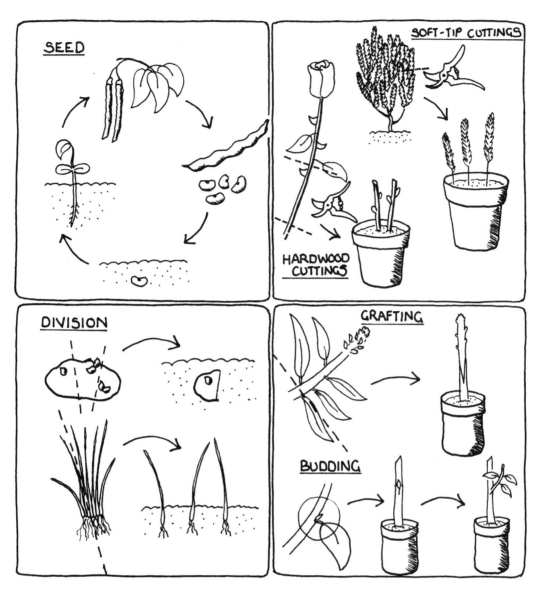

Figure 8.4 Propagation techniques.

market; they save money and grow vegetables all year. Saved seed can be kept at home, swapped with friends or it can be sent to a seedbank.

Problems with buying international seed

Today most of the world's seed is owned, grown and controlled by very big oil companies like Shell, or pharmaceutical companies like Monsanto's Pharmacia, and others such as Aventis and Syngenta. These companies grow seed as a commodity and not a food resource, to make a profit. They produce seed that is not the same as

local seed and which often exposes farmers to some of the following problems:

● Some seeds, called hybrids, won't grow well unless the farmers buy 'Seed + Fertiliser + Insecticide'. If they don't, then the plants grow badly and farmers can lose their crop. And when the crop fails because of drought, cyclone or flood, farmers sink into debt because they had to borrow money to buy all of the seed company's products.

● The big international seed companies want farmers to buy new seed from them every year, so they alter the plant seed genes so the

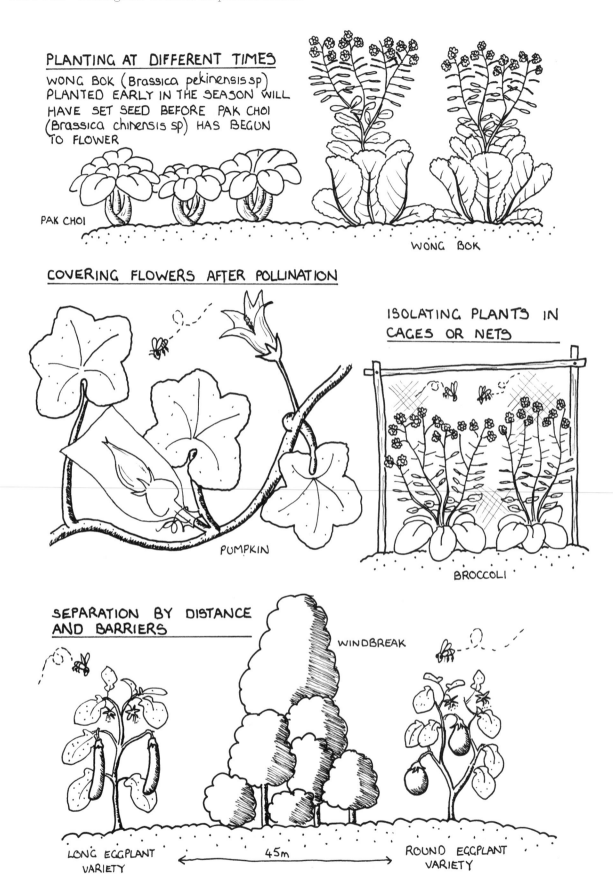

PLANTING AT DIFFERENT TIMES

WONG BOK (Brassica pekinensis sp)
PLANTED EARLY IN THE SEASON WILL
HAVE SET SEED BEFORE PAK CHOI
(Brassica chinensis sp) HAS BEGUN
TO FLOWER

PAK CHOI

WONG BOK

COVERING FLOWERS AFTER POLLINATION

PUMPKIN

ISOLATING PLANTS IN
CAGES OR NETS

BROCCOLI

SEPARATION BY DISTANCE
AND BARRIERS

WINDBREAK

LONG EGGPLANT
VARIETY

45m

ROUND EGGPLANT
VARIETY

Figure 8.5 Methods to keep varieties pure.

Figure 8.6 *Tagging selected plants and fruit.*

crops will not grow viable seed. These seeds produce sterile seeds in the next generation, which means they will not grow at all.

- Companies sell seeds that are grown far from where farmers live. For example, wheat seed developed and grown in dry areas is now sold in wet areas and gives low yields.
- Quality deteriorates when seed is too old, diseased, has low germination rates or it is not true to the original seed.
- Seed companies like to have only a small number of varieties, which are high yielding but have lost valuable genes for qualities such as disease resistance, tolerance of drought or flood, flavour and nutrition.

Keeping varieties pure

Ideally, you want your seeds to grow well and have the same desirable characteristics (which you choose) as the parents. This requires you to keep varieties 'pure'. Some plants are self-fertile and so

are automatically pure. To make sure you will have seed that is pure you can do the following:

- Grow vegetables at *different times* so the pollen cannot mix up. For example, choose both early and late season varieties.
- Grow vegetables of the same variety but grow them *some distance apart* so they cannot cross-pollinate. Also use *integrated planting* techniques, or windbreaks, to prevent the pollen being carried to another plant by insects or wind; for example, cucumber varieties, eggplant or tomato.
- Grow the same variety at the same time but *cover the flowers* (bagging) with paper bags to prevent the pollen from mixing with others.
- *Make baskets, cages or nets* to cover the whole tree or bush to keep it from air and insect pollination from other flowers.

Choosing plants for their seed

In general, select plants that are:

- heirloom varieties handed down from one generation to another
- local varieties grown as long as local people can remember
- varieties taken off the market that cannot be bought any more
- good recent arrivals.

Observe the plants and their fruits very carefully while they are growing so that you can determine which have special traits. A plant may be very good if it:

- survives in drought times
- it is heavy yielding
- has early-maturing fruit or late leaf and root crops
- is good in a special soil, whether clay or sandy or acid
- tastes delicious
- survives in a flood
- bears well in hot or cold seasons
- has large fruit or seeds
- is nutritious.

Usually you select a plant for no more than three of these traits. When you select the best plant for the reasons you have chosen, tie a coloured ribbon around it to remind yourself not to eat or pick the flowers, fruits or roots.

When to collect seed

- Collect the seed in the morning, before 10 o'clock but after the dew has gone from the plant. Collect from a part of the plant that is sunny and healthy, without diseases, insect attacks or eggs on it.
- Collect all fruits and vegetables when ripe or over-ripe. For chilli and capsicum, collect the seed when the outside skin is soft.
- For herbs, see that the seed is very ripe, pull the stem and root from the soil and hang the whole plant upside down in a cool dry place. Cover with a paper bag so that seed is not scattered or lost, and keep the stem dry.
- Collect seedpods of beans and cabbages when the outside skin is quite dry and full of seed.
- For vegetables with roots, make sure the fruit and seed are very ripe and collect the root and stem—as for herbs.

Figure 8.7 Traditional methods of winnowing and sieving seeds to remove chaff.

Cleaning, testing and drying seed

Seeds must be thoroughly cleaned before they are stored:

1. Winnow the seeds by shaking in an open-weave basket or sieve. Good clean seed is left after any sticks, stones, husks and dead seeds have separated out.
2. Alternatively, you can place the seeds in water. Dead seeds and dry plant materials will float, leaving the good seed behind.

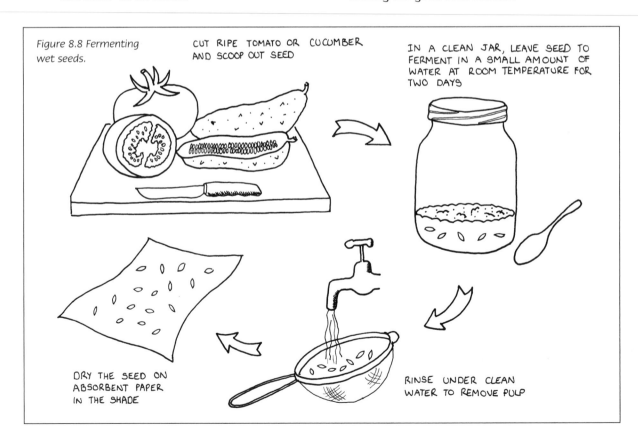

Figure 8.8 Fermenting wet seeds.

CUT RIPE TOMATO OR CUCUMBER AND SCOOP OUT SEED

IN A CLEAN JAR, LEAVE SEED TO FERMENT IN A SMALL AMOUNT OF WATER AT ROOM TEMPERATURE FOR TWO DAYS

DRY THE SEED ON ABSORBENT PAPER IN THE SHADE

RINSE UNDER CLEAN WATER TO REMOVE PULP

Figure 8.9 Storing seed correctly.

CLEAN, DRY SEED IS STORED IN PAPER OR PLASTIC BAGS IN CLEAN GLASS JARS WITH AIR-TIGHT LIDS

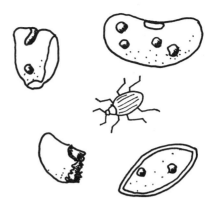

CHECK SEED REGULARLY FOR INSECT DAMAGE

3. For wet seeds, such as tomatoes, cucumber and rockmelon, ferment the seeds by leaving them in a small amount of water at room temperature for two days and then rinse well until all the pulp has gone. Dry the seeds on non-sticky paper.

Seeds are heavier when they are alive—dead seeds or seeds that insects have eaten inside are light and float in water. To test large seeds, place them in a glass of water: the living seeds will drop to the bottom and the dead seeds float to the top. If more seeds float than sink, select another batch of seeds and test them.

You may want to treat seeds to control diseases such as blackspot, blackleg and black rot. For largish dry seeds such as spinach and cabbage, place them in hot water at 50°C for 25 minutes then continue with the drying process.

Seeds must be quite dry before they are stored or they can rot from fungus, attract pests, or get diseases from virus or bacteria. However, while it is important to dry the seed thoroughly, be careful not to make it so dry as to kill it. Seed is properly dry when you cannot dent the seedcoat with your thumbnail or you do not leave a tooth impression when you bite it. There are several ways to dry seed:

- Spread the seed evenly on newspaper and place it out of the wind and hot sun. On a windowsill out of the sun is a good place or you could place the seed on flywire and turn it regularly.
- Place it in paper bags and hang them in a breezy spot.
- In wet weather place seed above a fire or heater but never at a temperature higher than 45°C.

Sorting dry seed
- When the seed is dry, shake it into an open-weave basket or sieve and let all the broken seed drop through.
- Store the biggest and cleanest seeds in a paper bag, write their name on the bag and record the details in your diary immediately.

Recording your collected seed
It is important to record the details about the seed you have collected because:
- you may forget why you saved it by the time you want to plant it
- you may want to give it to someone
- you may want to compare it with another variety.

Keep a notebook or old diary and write down the following details about when you picked the seed:
- name of vegetable
- special qualities of vegetable—disease resistant, long yielding

- dates of collection
- if there were special conditions at the time it was growing, such as very dry
- if you sent some seed to your seedbank, include these details on the packet:
    Address of sender
    Name of seed and special type
    Details of the seed—e.g. can grow in dry season
    Date of collection
- on the envelope you have placed the seed in, write the same details.

Storing and keeping seed

If seed is not stored well it can die, rot or germinate before you plant it. Seed can be kept alive for quite a long time if it is stored properly.

- Store the seed in plastic or paper bags inside glass jars or bottles that are airtight. This is very important because with no air any insects will die and fungi, viruses and bacteria have trouble surviving.
- You need a dark and cool place with a temperature between 5–20°C.
- Seed can be stored in a 1-metre-deep hole in the ground, under the house or verandah, but not outside.
- All seed will keep stored in a refrigerator for three to four years.
- If seed is stored for one year it is good to put it in a refrigerator for two or three days before sowing it.
- Check jars every two or three months for fungal growth or weevil eggs which may have hatched.

Remember, however, that seeds, like all living things, will eventually die so it is no use keeping them too long in storage. Also, different seeds have different life expectancies. Seeds are much better preserved by continually growing them ... and sharing them.

What happens at the seedbank?

A seedbank is a small office and store, or someone's home with a small- to medium-sized garden.

Usually the seedbank people really love and value seed. The seedbank:

- receives seed and does germination tests on all types of seed
- records the performance and conditions of growth and resistance to pests and diseases on the seed packet and in their records
- keeps very good, clear records about each type of seed
- stores some seed for short periods of time, such as until the next growing season
- trials, through growing, some new or very different or difficult seeds.

The seedbank does not:

- store very much seed—it sends seed out to growers or farmers
- charge for the seed
- buy seed—volunteers send seed to the bank
- grow very much seed—special farmers and gardeners are used as seed growers
- sell seed to multinational companies or others.

 Try these:

1. Find two plants that you would like to grow. Write a description of them using the table on type, traits, functions, etc. Mark on your home plan where you think they would grow well.

2. Find out the names of local vegetable and fruit cultivars that grow especially well near you. Can you fit one more in your garden, or get a friend to grow them?

3. Identify old neglected fruit trees or old-fashioned vegetables someone is growing close to you. These are called heirloom varieties and to be kept safe in your local area must be grown, not kept in seed packets. If you have a local area map then mark them on it, and if they are threatened take cuttings very quickly and continue growing them.

PART THREE

Applying permaculture

Using your experience and observations of soils, water, climate and microclimates, vegetation and windbreaks, you have analysed your site and now have a comprehensive view of what is on your land. All this information is baseline data, which will enable you to measure over time how effective your design is, once implemented.

In the next chapter you will read about three more design methods that will enable you to apply in detail the permaculture zones and place them on your final plan. Your resulting design will be as effective as the information you have gathered.

For each permaculture zone, and house, office, shop or factory and its land, specific aims and strategies are required to achieve sustainability. On some sites not all zones are possible due to size constraints or lack of resources. Your final plan will show the zones appropriate to your site and these will form a natural integrated design that can be tested against the ethics and principles of permaculture.

As you work through each chapter in this part and complete its tasks you will be building up a comprehensive land plan. By the time you complete Zone V you will have a design for every part of your land and buildings. The process is much like fitting pieces into a jigsaw puzzle. How well you do this will depend on your problem-solving ability and creativity.

Your main task for Zones 0 to V is to completely redesign your buildings and land using all your knowledge and skills.

CHAPTER 9

Reading your land

In my experience it is only when people use their senses to thoughtfully touch, smell, look, taste and record that they become fully aware of their surroundings. Often people who have lived in one place for years are surprised and say, 'I had no idea that tree was there … or that soil so good … or so dry', when they start to use all their senses fully.

It probably takes one or more lifetimes to get to know any land intimately; yet, by developing observational skills and environmental consciousness you will begin to learn what to do and when, while gaining pleasure from the exercise. This chapter is about close observation. Good, close observation leads to more accurate understanding of problems, assessment of potential and their resolution. In the initial stages, record everything in order to build up an inventory. This is simply a record of information about what is on your land, or in your house or office, and how these elements interact. Later you will use your inventory to draw up a site analysis. It will all contribute in some way to your final permaculture design.

Know your land intimately in all its moods: know what is there and what happens on and around it. Let the land shape you. The Earth is littered with mistakes, such as wrongly placed dams, houses, crops and even townships. Time spent now in getting to know your land is your best investment. You should spend as much time in looking, wondering and protracted thought as with a pencil and paper.

Our ethical task is to:
- hone, integrate and apply our observations
- consider thoughtfully all elements and their potential interactions
- refrain from early value judgments—for example, that aspect is 'too hot' or 'the slope is too steep'—until the inventory is complete.

Our design aims for taking an inventory are to:
- observe, collect and record everything, on- and off-site, about our land.

If we don't have design aims for taking an inventory:
- we tend to make speedy conclusions
- we lack detailed land knowledge
- we sometimes inappropriately transfer knowledge from some other place, which results in degraded land, costs money, destroys motivation and magnifies errors over time.

Ecological function of an inventory and site analysis

An important function of your work is to make connections between different variables and to see interacting patterns across the land among such elements as water, weeds, frost, wind and even diseases. Making these connections is basic to good design.

Routine inventories are made of Earth's resources by satellites and by cartographers; however, these don't usually tell you what the origins of problems are. They don't replace good personal observation but they do complement it.

Techniques and strategies for reading your land

When you start observing you may feel there is not much to notice. Observation is a skill that grows and grows. Eventually you will feel that if you were confined to a small garden for the rest of your life you could never know everything, nor ever be bored.

Your inventory and later site analysis together form the basis for your design. To get the whole picture you need two types of information: on-site and off-site.

On-site information

This is information you obtain from your own land. It consists of taking samples of plants and soils, measurements and experience of damp and dry patches, and shady and sunny aspects, and making notebook observations. How reliable these are depends on your skill and accuracy. Only a few individuals naturally make good observers. However, I believe that everyone has the potential to be sensitive and connected to what is happening on their land and improve their observations.

The strategy which is most important for observation and deduction is to walk around your land frequently at different times of the day and in different seasons to see the multitude of things occurring and draw connections between them.

In addition, there are several skills you will need to foster.

Targeted observation

This is carried out on the actual site and with a particular theme or problem in mind. It could be a problem such as weed encroachment or soil erosion. Your observations are the careful noting of anything and everything that may be connected with the problem. If you've had a weed problem then you might look at animal propagators, soil and water enrichment, and wind direction. Your initial list may be quite long and you will follow up and verify each item on the list. Sometimes this entails touching, smelling and tasting. Next, cross off any improbable or unlikely items when you are sure you don't need them. Finally, you will have a short list and you can test each of these as a hypothesis for the cause or remedy of the problem.

Deduction

In permaculture terms, deduction is examining another landscape similar to your own to find a design solution. If you experience severe cold winds and you are trying to design good windbreaks, then look for a similar site where the windbreaks—either naturally or humanly designed—are effective, and then copy these, incorporating whatever changes you need for the windbreak to work for you. Nature is rich in examples and is a good mentor (see Figure 9.1).

Reading patterns

This involves making connections between your observations and deductions. If a breeze comes up every evening about five o'clock then you may see this as a beneficial summer phenomenon. Perhaps mists will drift in at about sunset every day and so you do not have to water as much, even though it does not rain. Other patterns can be those of place: you notice which groups of plants grow well together, or in certain aspects (such as liking

Figure 9.1 Copying examples in nature. Try this with productive species.

morning sun). Patterns exist in time, in place and in relationships. One very obvious pattern is that of the daily and seasonal movements of the sun. If you know at what time and in what season part of a garden will be in shade, then you can successfully select plants to grow there.

Experiential understanding

This is what you know about your land from your own experience of being there. Sometimes you cannot even express your experience clearly. For example, you may know very well that one place is just not good for planting. You may not know why. However, sometimes you find out later. This is partially a 'gut' feeling and it should be trusted.

Analysis

Analysis is a very good method for placing a new element accurately into a landscape and ensuring that it largely meets its own needs. It is particularly useful for placing animals into a system. When you want to introduce a new animal into your design, you list all its *yields* and its *needs*. Some of the yields can be behavioural, such as scratching the ground, pollinating, etc., while others will be produce like eggs and meat. The animal's needs for food and shelter are designed into the system so the animal requires as little human maintenance as possible (minimal human inputs). For chickens you will design grains, greens and medicines to grow where the chickens run. You will encounter this method again when you study poultry and bees in Zone II and larger animals in Zone III (farming).

Mapping

This is the method you have been using to record your observations and findings on paper. You can start with a rough sketch of all the features in relation to each other or you can draw up a scale plan. If your site is quite small, put the plan on A4 paper; however, if it is a large farm or neighbourhood then you can use A1. Remember, at this stage rough drawings are fine. Unless you are aiming to become an expert designer, scale is not vitally important.

Using a strong black pen, draw your base plan which shows the main permanent features such as roads, buildings, dams, rivers, etc. Place tracing paper over the top for your inventory and site analysis (see Figures 9.2 and 9.3 of Rob's place).

If you have a lot of information, use several clear sheets over the top of your base plan. For example, if you feel your inventory is getting too cluttered with paths and buildings then you can draw all your plants, or water, or soils on a separate clear sheet. Your base plan and your inventory are now integrated as one presentation.

Later, in your site analysis, you will include climates and microclimates, aspects, views, state of soils, limiting factors and so on from Part Two.

Off-site information

As implied, this is data you collect from sources off your land. Generally, you do not need to find original data for yourself because in most cases the information you want already exists somewhere. Look in the phone book and start with government departments. Remember to investigate your local

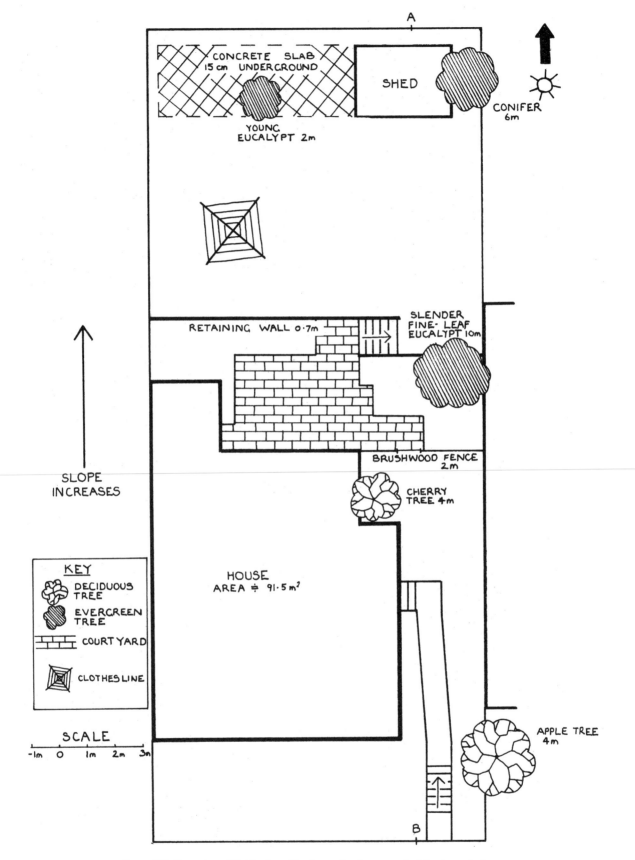

A

CONCRETE SLAB
15 cm UNDERGROUND

SHED

CONIFER
6m

YOUNG
EUCALYPT 2m

SLENDER
FINE-LEAF
EUCALYPT 10m

RETAINING WALL 0·7m

BRUSHWOOD FENCE
2m

SLOPE
INCREASES

CHERRY
TREE 4m

KEY

DECIDUOUS
TREE

EVERGREEN
TREE

COURT YARD

CLOTHES LINE

HOUSE
AREA ≑ 91·5 m²

SCALE

-1m   O   1m   2m   3m

APPLE TREE
4m

B

*Figure 9.2 Base plan of Rob's place. The base plan is a record of the boundaries
and existing features of the site, and is the first stage in the design process.*

garden and historical societies, visit the local library, and speak to elderly residents, all wonderful sources of information.

Climate data

In country areas the post office collects basic information on rainfall, wind, evaporation, frost and temperature for the local area. The weather bureau has much more comprehensive data over a long period. Also, your local government office can sometimes give you weather statistics for rainfall, its incidence and distribution; wind velocity, direction and strength; sunny days and cloud cover; frost and mist. These days most weather data include pollution readings for air quality.

Maps

The Lands Department or Central Mapping Authority supplies a large range of maps in different scales and for different purposes, such as vegetation, soils, contours, photographic maps, land use. Local government offices also have maps on industrial and environmental zoning and planning.

Cadastral maps are accurately surveyed maps of land giving the boundaries to scale. Local government uses them to calculate your land rates.

Topographic maps are very useful because they have contour lines marked on them.

Vegetation

You can find maps, plant lists and people to help you identify indigenous plants, weeds, food, and rare plant species from a range of sources. National Parks

and Wildlife departments, universities, libraries, local government, botanic gardens, conservation societies and local garden clubs will all help you. Organisations dedicated to bushcare and wild plant rescue also have people with good historical knowledge of local areas and plant communities.

Water

The Department of Water Resources or water boards will assist with information and structural and legal details on water quality, building dams, sinking bores, or diverting or protecting streams. In addition, community organisations such as Stream- or Coastwatch have people with good local information and memories.

Rural affairs and industries

The departments of Agriculture and Rural Development assist with advice on horticulture, animal husbandry, revegetation, soils, salinity and general landcare issues.

Planning and legislation

If you need information on environmental planning issues, including legislation dealing with endangered species, roads, access, local and regional environmental plans, heritage orders and water control, then contact the appropriate departments of Planning and the Environment.

Putting it all together

We have two basic resources: people and land. Throughout this book they become ever more

Figure 9.3 Profile of Rob's place, showing the slope of the land along the north-south boundaries.

tightly interwoven. Your skills in observation, analysis, correlation and reading patterns are developing through the 'Try these' exercises. Your knowledge and experience of the land is developing through systematic study of the ecological themes in Part Two. Now in Part Three you will have your land fully designed by imposing patterns, zones and sectors, and you will be able to evaluate it for:

- sustainability of soils, water and species to satisfy the intergenerational equity principle (see Chapter 3, page 19)
- high yields
- diversity
- low maintenance at maturity (15 years).

Try these:

1. Take a land inventory. *Walk around your property; don't do a windscreen appraisal!* You can easily measure a suburban block or small farm. For a bigger site, take the length of the boundaries from a map or deposited plan, obtainable from the local council or government body.

- Take your tape measure (a 30-metre one is best) or a string knotted in 10-metre lengths. Now walk, and measure, the boundaries of your land. Look over fences and record on your sketch plan everything you see, even if it isn't on your land. Show everything in your sketchbook. Rough sketching is okay; it doesn't have to be completely accurate.
- Take bags or jars with you for soil and plant samples. On your plan show all the soils, water vegetation and land use.
- Show roads, slopes, aspects and land health, such as eroded, good grass cover. Be as detailed as possible.
- Stick your sketch up somewhere at home and stare thoughtfully at it

very often. Try to see relationships between soil and water, or soil and plants, or land use and plants; for example, 'There seems to be seepage at the bottom of the slope by the look of the weeds there.' Then go and look again regularly. Make sure you choose different weather: a fine clear day and a cold rainy one. Revise your observations as you learn more.

2. Start now to make preparations for your site analysis. Draw on a new piece of paper a second outline of the main permanent features of your land and include buildings, fences, creeks, paths, roads, water tanks, taps and contours. You may be able to obtain a cadastral map and a topographic map from your local government office and take your boundaries off it.

You will be working on this plan and adding a lot more detail as you read through the next few chapters. Stick it up beside your inventory and compare both. Just do what you can—there is no good or bad. Whatever you put into this exercise you will be the richer for it. Invest well.

CHAPTER 10

Developing your design methods

Zones and *sectors* are names given to the principal design methods of permaculture used for applying information and understanding to a design. You can apply them to new 'bare' land or to land that is already in use but which you want to make more sustainable, perhaps by using organic methods or modifying existing enterprises. This is called retrofitting the land, or rolling permaculture.

Our ethical task is to:

- approach nature as a friend and ally whose ways must be understood and whose counsel is needed
- apply design methods that do no harm and respect the natural inherent qualities of the land.

Our design aims for our site are to:

- produce high yields
- require low maintenance
- produce a non-polluting lifestyle and enterprises
- approach a closed system that meets its own needs
- be diverse and resilient enough to endure adverse conditions.

If we don't have design aims:

- we may live in a toxic, vulnerable and low productivity landscape

- it will cost a lot of money, contribute to environmental degradation and diminish Earth's scarce resources
- we will be much less effectual in living a sustainable life.

Sector analysis

In this design method the four points of the globe—north, south, east and west—are observed and related to the movement of wind, sun and water, and microclimate factors This analysis begins by observing the factors originating outside the boundaries and, taking each aspect in turn, noting how renewable energies affect the entire property. Figure 10.1 shows a sector analysis for Rosie's farm, Figure 10.2 is Rob's sector analysis, and Figure 10.3 shows the sector analysis for my place. All sector analyses show:

- summer and winter sunrise and sunset
- warm gentle summer winds and harsh summer storms
- cold, gusty wind direction

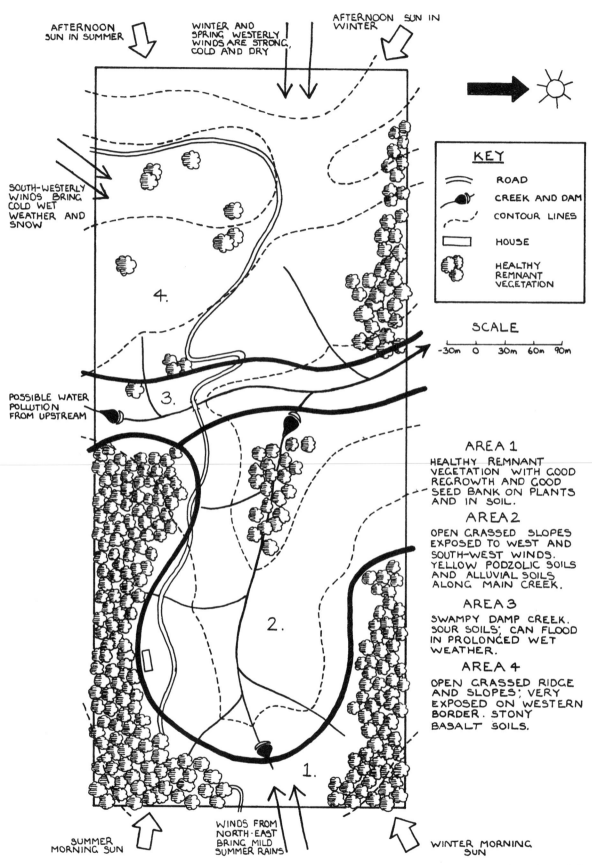

AFTERNOON
SUN IN SUMMER

WINTER AND
SPRING WESTERLY
WINDS ARE STRONG,
COLD AND DRY

AFTERNOON SUN IN
WINTER

SOUTH-WESTERLY
WINDS BRING
COLD WET
WEATHER AND
SNOW

KEY

ROAD

CREEK AND DAM

CONTOUR LINES

HOUSE

HEALTHY
REMNANT
VEGETATION

SCALE

-30m   0   30m   60m   90m

POSSIBLE WATER
POLLUTION
FROM UPSTREAM

3.

4.

2.

1.

AREA 1
HEALTHY REMNANT
VEGETATION WITH GOOD
REGROWTH AND GOOD
SEED BANK ON PLANTS
AND IN SOIL.

AREA 2
OPEN GRASSED SLOPES
EXPOSED TO WEST AND
SOUTH-WEST WINDS.
YELLOW PODZOLIC SOILS
AND ALLUVIAL SOILS
ALONG MAIN CREEK.

AREA 3
SWAMPY DAMP CREEK.
SOUR SOILS; CAN FLOOD
IN PROLONGED WET
WEATHER.

AREA 4
OPEN GRASSED RIDGE
AND SLOPES; VERY
EXPOSED ON WESTERN
BORDER. STONY
BASALT SOILS.

SUMMER
MORNING SUN

WINDS FROM
NORTH-EAST
BRING MILD
SUMMER RAINS

WINTER MORNING
SUN

10.1 Site analysis and sector analysis of Rosie's farm.

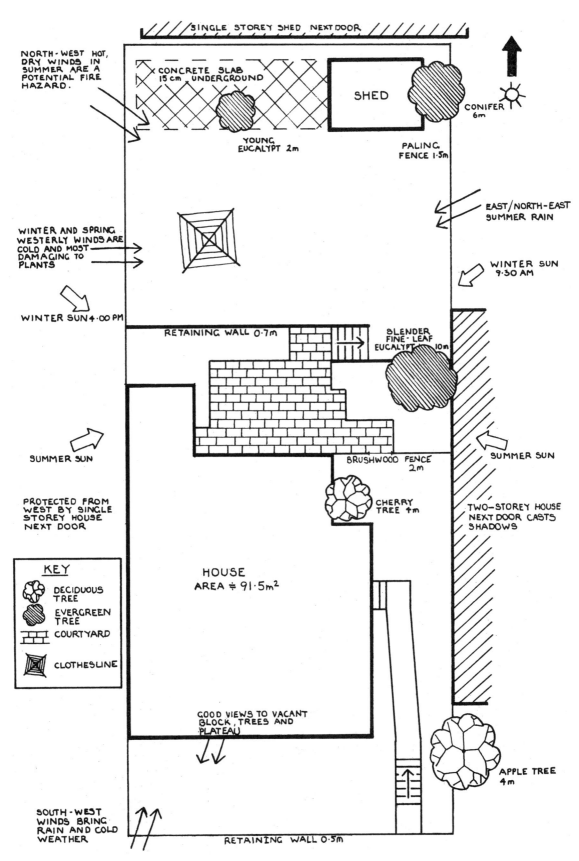

NORTH-WEST HOT, DRY WINDS IN SUMMER ARE A POTENTIAL FIRE HAZARD.

SINGLE STOREY SHED NEXT DOOR

CONCRETE SLAB 15 cm UNDERGROUND

SHED

CONIFER 6m

YOUNG EUCALYPT 2m

PALING FENCE 1·5m

EAST/NORTH-EAST SUMMER RAIN

WINTER AND SPRING WESTERLY WINDS ARE COLD AND MOST DAMAGING TO PLANTS

WINTER SUN 9·30 AM

WINTER SUN 4·00 PM

RETAINING WALL 0·7m

SLENDER FINE-LEAF EUCALYPT 10m

SUMMER SUN

SUMMER SUN

BRUSHWOOD FENCE 2m

PROTECTED FROM WEST BY SINGLE STOREY HOUSE NEXT DOOR

CHERRY TREE 4m

TWO-STOREY HOUSE NEXT DOOR CASTS SHADOWS

KEY

⬢ DECIDUOUS TREE

⬢ EVERGREEN TREE

▦ COURTYARD

▨ CLOTHESLINE

HOUSE AREA ≑ 91·5m²

GOOD VIEWS TO VACANT BLOCK, TREES AND PLATEAU

APPLE TREE 4m

SOUTH-WEST WINDS BRING RAIN AND COLD WEATHER

RETAINING WALL 0·5m

10.2 Sector analysis of Rob's place. The sector analysis shows the type and direction of wild energies (rain, wind, floods, fires, sun) which affect your property.

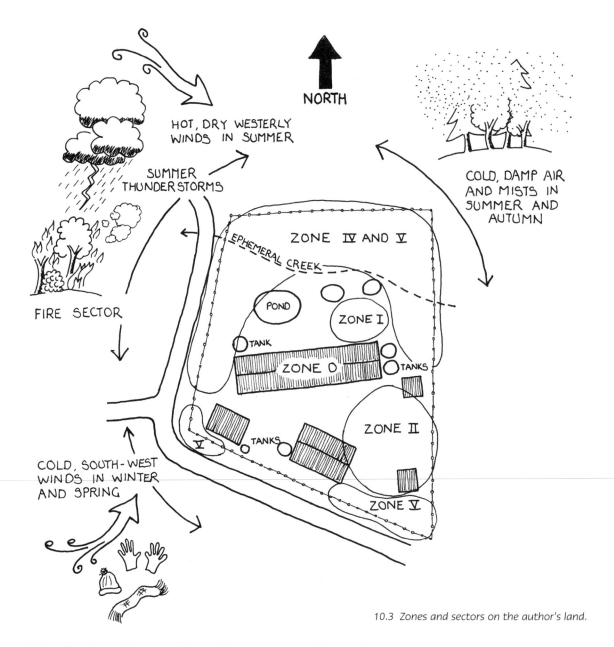

NORTH

HOT, DRY WESTERLY WINDS IN SUMMER

SUMMER THUNDERSTORMS

COLD, DAMP AIR AND MISTS IN SUMMER AND AUTUMN

ZONE IV AND V

EPHEMERAL CREEK

FIRE SECTOR

POND

ZONE I

TANK

ZONE 0

TANKS

ZONE II

TANKS

V

ZONE V

COLD, SOUTH-WEST WINDS IN WINTER AND SPRING

10.3 Zones and sectors on the author's land.

- likely direction of danger of fire, pollution, flood and cyclones
- quality of rain—heavy, mist, damaging—and frosts
- neighbouring vegetation and buildings and other threats and advantages.

By identifying the type, intensity and direction of external effects you can include them in your later design considerations so their impact can be minimised or maximised. For example, you may need to choose fine-leafed shrub species to filter out dust pollution. If you live in an area where fire, cyclone, nuclear hazard, floods or pollution are

problems, then these must be included.

Use coloured pencils or crayons when drawing your design as it makes your work easier to read. I use blue for water, brown for roads and buildings, black for fences and green for vegetation.

Zones

Zones assist designers to place enterprises on land to achieve minimum inputs, resource recycling, high yields, low maintenance and resilience. The number of useful connections made among the elements is critical to the sustainability of each zone. Zones can be thought of as a series of concentric rings, starting with the home centre as Zone 0 and working outwards. Your placement of plants, animals and structures in each zone depends on its yields, functions and maintenance requirements. For example, chickens are placed quite close to the house since they give eggs most days and require a constant supply of clean water, whereas an apricot tree, which yields all its crop in a few days of the year and needs less frequent watering and feeding, is further from the house. Zoning, the main design strategy of permaculture, yields a huge amount of diversity and biomass compared with a similar area of monoculture.

There are six zones, starting with Zone 0, your home, and moving out to Zone V, which is natural forest. Each zone has appropriate functions and enterprises. To decide the size and type of enterprise for each zone ask yourself the following questions:

- How productive is this zone?
- How much energy, water and nutrient resources will it require?
- How frequently will it need maintenance and attention?

Table 10.1 is a summary of the zones to help you think about landscape design in this way.

Most suburban blocks can contain Zones I and II, either separately or integrated. These two zones use your grey water from sinks and washing machines and provide most of your food (except grains) while recycling all your organic waste into quality humus. They are also fundamental to increasing your self-reliance while providing security from toxic chemical sprays, unprincipled agricultural businesses and inflated food miles.

Zones are initially imagined as concentric circles, with Zone I as the most productive and resource hungry, and Zone V as long-term, lower maintenance and lower resource use. It is important to remember these zones are simply concepts and are not fixed land boundaries. Zones flow in and out of each other and are flexible, changing over time if necessary. Zones fit land according to its inherent potential and qualities. They reflect slope, orientation of the sun and the land's natural

TABLE 10.1: SUMMARY OF ZONES

Zone	Summary of function and placement
Zone 0	The home, office, factory or shop is a voracious consumer of materials, water and energy and usually generates waste into the environment. Sustainable design of Zone 0 is a priority.
Zone I	The home kitchen, or vegetable, garden is close to the house, high yielding, intensively cultivated, and produces mostly herbs and vegetables.
Zone II	This is an intensively cultivated, heavily mulched and closely planted orchard with grafted and selected fruit trees. You might visit it once a day and probably run smaller animals such as ducks and geese. (This isn't essential.)
Zone III	This is the farming zone and requires less maintenance than Zones I and II. Commercial crops are grown, the plant species are hardier, and animal forage systems are used. It may be an organic orchard, nut forest, or extensive organic poultry system. You may grow cereal and industrial crops, or intensively farm beef, dairy or sheep, or raise deer and goats. It is protected by multifunctional windbreaks and alley crops. It is usually further from Zones I and II and before Zone IV.
Zone IV	These are harvest forests, well timbered and for long-term development. Tree species are harvested sustainably for building, mulching, firewood and precious timbers and this zone can carry complementary grazing animals such as cattle, deer and pigs or indigenous animals at low stocking rates.
Zone V	This is the indigenous conservation zone providing protection for soils, water, air and species richness of indigenous plants of the region. It functions as a reserve, a regrowth area and a bank to stock wildlife in the future. If possible, it is connected to a national park or reserves through wildlife corridors.

features. For example, you do not remove indigenous forests to plant an exotic orchard. The natural bushland is incorporated into wildlife corridors or Zone V. In addition, as you know, steep slopes, natural ecosystems, rivers and springs are more resistant to damage and degradation if left under natural vegetation. So the five concentric circles quickly get teased into patterns that do not readily align themselves with this model. Figure 10.4 shows how zones appear when you allow for such factors. The house and Zones I and II sit in a protected clearing bathed in sunlight.

Every site is different, so when you are designing your site apply conscious, thoughtful and protracted observation to discover how zones and sectors can be applied to increase yields, decrease work and move along the line of increasing sustainability.

Relative location

This is another design method often used in permaculture that works together with zones and sectors to maximise your land's potential. Relative location enables you to place elements such as plants, structures and animals in relation to each other so they enhance each other's functions. You will require this skill to achieve high yields and reduce energy and water use. For example, tank water is stored close to bathrooms and kitchens and the grey water is cleaned close to gardens where it is used. To reduce work, the garden shed is placed halfway along the block because tools are used uphill and downhill from it. The chickens are on one of the highest points so that carrying their manure downhill is easy, and the nutrient moves downhill to fertilise the fruit trees when it rains.

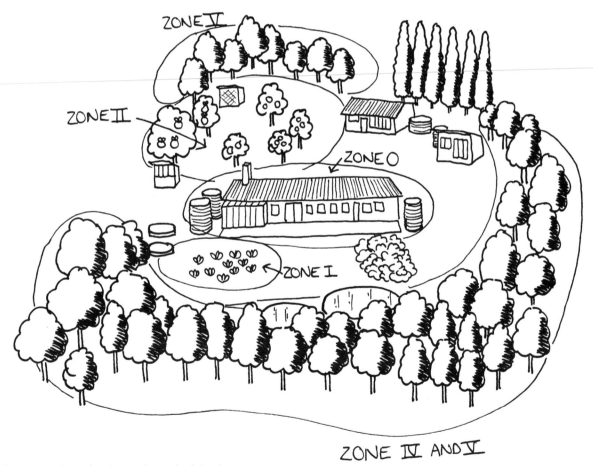

Figure 10.4 Zone planning at the author's land.

You have now been introduced to the main permaculture design methods. You have also drawn whole site water and soil plans. Now refer back to the Whole Site Water Plan in Chapter 4 (see page 36) and see how it looks placed beside the zones in Figure 10.3 in this chapter. Look again at how the site analysis and sector analysis have been combined for Rosie's place in Figure 10.1.

Remember these guidelines to help you monitor your design:

- Start small and get it right.
- Make incremental changes to an existing system.
- Keep measuring your ecological footprint.
- Move to greater and greater sustainability.
- Work to a time, budget and plan.
- Set measurable objectives and monitor them.
- Change your objectives if you think of something better or unexpected with favourable outcomes.

 Try these:

1. Check and finalise your site analysis. Look at the site analysis of Rob's place in Figure 10.2. You can see how it interprets the land for him. Then you can look at Rosie's farm (Figure 10.1) for an analysis of a larger site. Have you included climates, microclimates, aspects, views, soils, limiting factors and even history? Has your site been mined for soil minerals, for example? Stick your site analysis up where you can see it and keep working on it as you make new observations. (It can never be perfect or complete because this would take a lifetime. There's a lifetime's study in every piece of land.) What was surprising to you when you looked at your land very closely? Do you consider your observational skills have improved?

2. Prepare for your final design. Get a new large sheet of paper and draw the main permanent features on it. Do a sector analysis of your land on this sheet, writing your observations outside the boundaries. Look again at Rob's place and Rosie's farm.

CHAPTER 11
How and where we live: Zone 0

Your home, Zone 0, is the place where you spend most of your time, money and other resources. It is important that your home contributes to your health and wellbeing and is not a drain on environmental resources or your finances.

Permaculture is not only about living things; it is also about how we live and what we use. It is about our homes and the quality and cost of the lives we carry on in them. In permaculture we aim to design new houses or to retrofit old ones so they are comfortable living places that preserve and restore the Earth's resources.

Energy supplies from oil and gas are fast diminishing. In fact, we are considered to have entered an age of energy descent for oil, and with wars and natural disasters even the reserve supplies are diminished. No one can expect that the same amount of relatively cheap energy will be available in the future unless solar energy is harnessed to run transport and industry. (At present, solar energy units have a cradle-to-grave energy cost which makes them unsustainable; however, efficiency in this technology is improving rapidly and the costs are falling.) Water supplies are even more gravely threatened.

We need to practise the precautionary principle and change our buildings, lifestyle and consumer use to live comfortably and simply with less. When the resources become very scarce, the means may not be available to make the necessary changes. Anyone who does not alter their life to live better

and simply now, while resources are still relatively cheap and available, may not be able to later because of scarcity of materials or elevated prices. In practice this means:

- reducing inputs into homes and constantly reducing waste
- ensuring that outputs can be contained on site, are non-polluting and enhance other functions
- living and working without waste
- increasing and replacing non-renewable resources.

 Our ethical task is to:
- build new homes and retrofit old ones using renewable materials, low energy and minimum pollution
- live in our homes with simplicity.

 Our design aims for homes are to:
- match space and function
- admit and store the sun's energy when needed, and remove and exclude heat when not required

- be simple, honest and economic with resources
- use living processes to recycle waste.

If we don't have design aims our homes:

- will cost us excessively to build, to live in and repair
- may be polluting and bad for our health
- will use and waste non-renewable resources
- will not be easy to live in
- may prove very expensive in the future.

Live differently, live well

For thousands of years there have been cultures that have enjoyed a high quality of life without compromising the needs of present and future generations by destroying or polluting life processes. Compare this with modern Western lifestyles which are characterised by:

- excessive consumerism
- alienation from nature
- over-processing of materials
- over-consumption and dissatisfaction
- embarrassing levels of waste of all resources
- overly busy and complicated lives.

In affluent countries most homes, offices and shops can be thought of as having one or a combination of the characteristics examined in Table 11.1.

All these houses take in clean good materials and spew out air, soil, water pollution, organic-putrible and foul materials (see Figure 11.1). So how can you live differently and still live well? You can build a new house or retrofit an old one. And you can make significant changes to your lifestyle and buildings that will reduce the destructive impact that our living has on the environment. Changes in your life are normally accompanied by improvement in the quality of your life; that is, in time, health and money.

Siting a new house

The difference between comfort or misery and great expense can depend on siting your house correctly. So consider the following factors before you select your land. Look at Figure 11.2 as we discuss the important factors.

Climate

- In hot climates choose a site with cooling breezes and shade.
- In cool climates choose a site with sunny aspects and protection from cold winds.

TABLE 11.1: HOUSE CHARACTERISTICS

House type	Characterised by
Consumer junkies	These houses devour huge amounts of finite resources and then release toxic or polluted air, water, or materials into the environment. In most Western countries the appetite to fuel this pollution stream seems insatiable with consumption of energy, water and materials increasing every year. This is increasingly copied by wealthy people in poor countries.
Sick houses	These homes have problems caused by artificial chemicals or processed materials that have been used in the construction and furnishings, such as asbestos, glues. In addition there is the problem of low-level electromagnetic radiation which emanates from the large number of electric items used in homes.
Vulnerable houses	These are the houses almost or completely dependent on vital resources such as water or energy from one source only. Most people do not have another water alternative. Also, many homes are totally dependent on one energy source which could be corrupted in some way, and Earth's energy resources are running out. In many homes the only source of food is a supermarket.

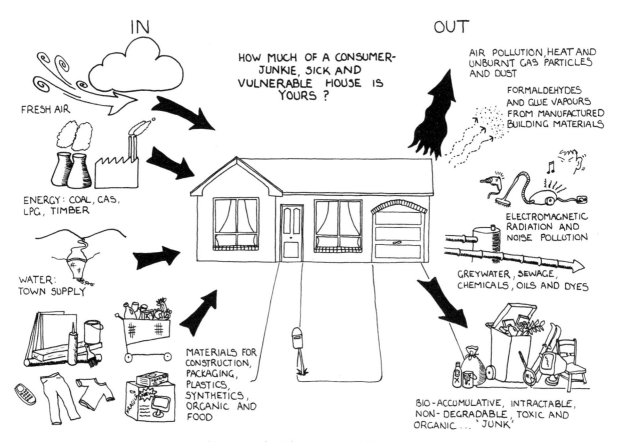

IN

OUT

HOW MUCH OF A CONSUMER-JUNKIE, SICK AND VULNERABLE HOUSE IS YOURS?

FRESH AIR

ENERGY: COAL, GAS, LPG, TIMBER

WATER: TOWN SUPPLY

MATERIALS FOR CONSTRUCTION, PACKAGING, PLASTICS, SYNTHETICS, ORGANIC AND FOOD

AIR POLLUTION, HEAT AND UNBURNT GAS PARTICLES AND DUST

FORMALDEHYDES AND GLUE VAPOURS FROM MANUFACTURED BUILDING MATERIALS

ELECTROMAGNETIC RADIATION AND NOISE POLLUTION

GREYWATER, SEWAGE, CHEMICALS, OILS AND DYES

BIO-ACCUMULATIVE, INTRACTABLE, NON-DEGRADABLE, TOXIC AND ORGANIC... 'JUNK'

Figure 11.1 The sick, consumer-junkie house.

Topography

- You have greater control over soil and water on a site with a slope not greater than 15 degrees.
- Western slopes are often very hot and dry, and polar-facing slopes can receive freezing winds.
- Hills behind your site block severe winds and assist in capturing surface water.

Water

As you read in Chapter 4, water is a primary selection factor for land. When it's raining hard, observe where the water goes in order to decide how you could work with it or capture some for future use. Dams are expensive to build, but if you don't have other water sources, they are a priority.

In drier areas, your land potential will be determined by rainfall. As a crude guide, 80 centimetres (30 inches) can be considered minimal

to maintain a reasonable standard of living and support enterprises. However, consider rainfall distribution because 80 centimetres falling in three months also means nine months of drought. You will need to collect enough water in your rainy season for the dry months. Remember, try to live and farm within your rainfall budget and alter your storage capacity accordingly.

Practise Keyline Water Harvesting to place clean-water dams uphill and use gravity to distribute water downhill (see page 41). Grey-water and aquaculture dams are placed downhill where the water is cleaned before being released to local creeks.

Bore water, traditionally used to bolster water supplies in dry areas, is an increasingly unreliable water source. It is often saline, or alkaline, and increasingly polluted by excessive chemicals from farm run-off. Over-used and wasted in the past, bore water is likely to be metered and charged for highly in the future.

Soil

Clay soils can shrink when dry and swell when wet, causing houses to crack. However, some clay is useful if you want to use mudbricks, cob, pise or rammed earth. Very sandy or shale soils won't hold water well in dams.

Surrounding land use

Check with your local council for their environmental plans. It is not much fun to build the house of your dreams only to find that a powerhouse, major highway or chemical plant will be built next door.

Access

Access can be very expensive if the council tells you that you must build the access road. You may need to consider owning a four-wheel-drive vehicle if you wish to avoid swimming across rivers. Bridges may

also need to be replaced. Access roads built along contours act as swales and require less maintenance.

Vegetation

Leave all remaining native vegetation and work around it. Don't remove any vegetation until you are ready to replace it. Soils hate to be left naked.

House orientation

How you orient your house on the land is a big factor in influencing the type and quantity of inputs needed to make your house comfortable all year round. In general, in temperate areas orient the long axis east–west, with the main daily living areas sited on the sunny side so as to benefit from winter sunlight.

House orientation changes with latitude and climate. So, for example, in high latitudes (in places

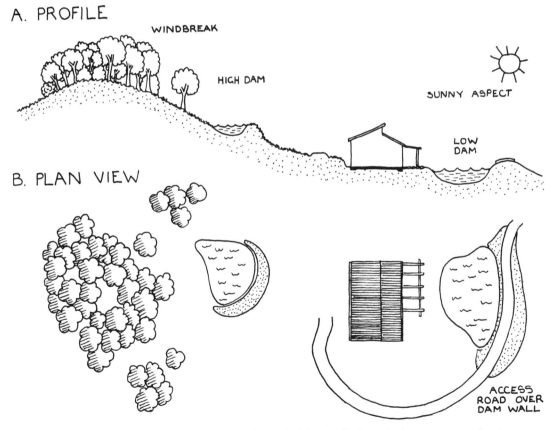

Figure 11.2 Site selection. This diagram shows the ideal profile for a site in a temperate climate. The trees on the ridge act as a windbreak and a recharge area for ground water; water can be gravity fed from the high dam to the house; the house is placed on the lower slope to receive maximum radiation; and the low dam reflects light to the house and modifies surrounding temperatures.

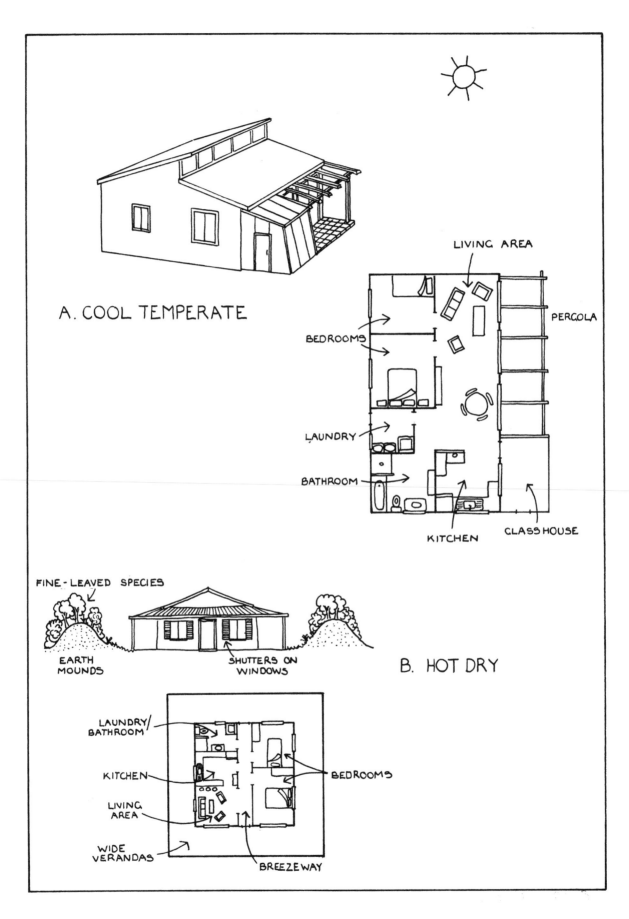

A. COOL TEMPERATE

LIVING AREA

PERGOLA

BEDROOMS

LAUNDRY

BATHROOM

KITCHEN

GLASS HOUSE

FINE-LEAVED SPECIES

EARTH MOUNDS

SHUTTERS ON WINDOWS

B. HOT DRY

LAUNDRY/ BATHROOM

KITCHEN

LIVING AREA

WIDE VERANDAS

BEDROOMS

BREEZEWAY

closer to the magnetic poles), orient buildings to receive more westerly sun. In desert areas and in lower latitudes, orient houses to minimise westerly sun. There are many books available that provide this information and an environmentally conscious architect can give you advice.

Figure 11.3 Examples of house designs suited to different climates.

House design

Once the right site is selected for your house, there are several factors to consider in its design.

Climate

Figure 11.3 demonstrates how you can design your house layout according to the climate and your requirements.

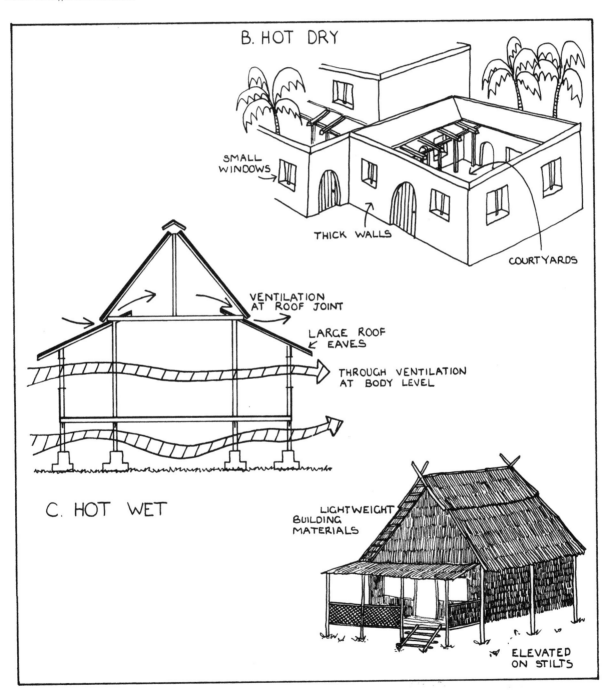

113

- In deserts and hot climates, buildings need wide shady verandahs with plenty of cross-ventilation.
- In cold climates, houses need exposed (sun-facing) glazing to heat and light the home.
- In hot wet climates, houses are better built on stilts to maximise evaporative breezes.

Building technology

Whether you are building a new home or retro-fitting one, materials and technology in buildings can be assessed for their ecological footprint, or their cradle-to-grave cost. Materials are great consumers of non-renewable resources, so use design criteria to test them.

Water supply, storage and use

In the previous chapters we have looked at collecting, storing and cleaning surface water. It makes good sense to be self-sufficient in water. Your water audit told you how much is available, how much you use and how much you can reduce, and which savings have the greatest impact.

Your water audit also allowed you to know how much profitable surplus you have to increase yields or productivity in other areas. For example, hot water can be channelled to tanks in a glasshouse to give extra warmth to plants, then siphoned off to a washing machine and finally to the garden. Eventually you will bring it back into the house as

an apple or cabbage, and no longer pollute rivers, lakes and oceans.

Energy usage

It is important to understand how you use energy in your home in order to see where you can make changes to achieve financial and resource economy. Hot water is obviously a major consumer of energy and so is heating in cool climates. To reduce heating costs, find out where heat is lost from your home. Figure 11.4 shows how energy is lost from a warm room. First you insulate your ceiling, then seal windows, walls and floors—in this order of priority.

It is vital that you also consider your method of heating. The Department of Minerals and Energy, or the equivalent government department where you live, puts out free annual information on the cost of useful heating. Use their information to achieve the greatest possible saving of money and resources.

The best technologies make use of almost infinitely renewable energy such as hydro, solar and wind. They can be used for heating, cooling, light and other useful work. Passive solar energy, which requires people to be active in controlling sunlight, is a very efficient form of heating. Essentially, light energy is stored in thermal mass as heat, which is radiated back after sunset.

As you make changes to reduce your energy consumption, keep monitoring your energy footprint. What are you costing our Earth?

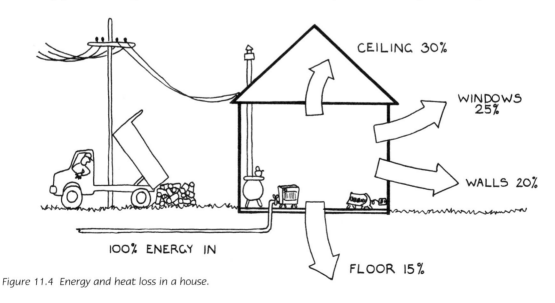

CEILING 30%

WINDOWS 25%

WALLS 20%

100% ENERGY IN

FLOOR 15%

Figure 11.4 Energy and heat loss in a house.

APPLIANCE	WATTS	DAY 1 MON 12/10	DAY 2 TUE 13/10	DAY 3 WED 14/10	DAY 4 THUR 15/10	DAY 5 FRI 16/10	DAY 6 SAT 17/10	DAY 7 SUN 18/10	AVERAGE DAILY USE	WATT HOURS
REFRIGERATOR	300	6 hrs	6 hrs	6 hrs	6 hrs	6 hrs	6 hrs	6 hrs	6 hrs	1800
ELECTRIC KETTLE	1800	/// (15 mins)	/// (15 mins)	/ (5 mins)	// (10 mins)	/// (15 mins)	##/ (25 mins)	//// (20 mins)	15 mins	450
STOVE/OVEN	2,400	30 mins	—	60 mins	45 mins	—	30 mins	45 mins	30 mins	1200
HOT WATER SYSTEM	4,800	1 hr	1 hr	1 hr	1hr	1 hr	1 hr	1 hr	1 hr	4800
TELEVISION	150	—	30 mins	—	—	2 hrs	2 hrs	2.5 hrs	1 hr	150
STEREO/RADIO	100	30 mins	1 hr	30 mins	1 hr	30 mins	4 hrs	3 hrs	1½ hrs	150
VACUUM CLEANER	500	—	—	—	—	—	35 mins	—	5 mins	42
WASHING MACHINE	485	—	—	—	—	1½ hrs	—	—	13 mins	120
POWER TOOLS	SAW 310 DRILL 430	—	—	—	—	—	15 mins 5 mins	—	3 mins	37
LIGHTS										
KITCHEN FLUORO	40	1½ hrs	2 hrs	1 hr	1 hr	2¾ hrs	2 hrs	1½ hrs	1½ hrs	60
LIVING ROOM	75	2 hrs	1 hr	1 hr	2 hrs	3 hrs	2 hrs	3 hrs	2 hrs	150
BEDROOM	75	30 mins	15 mins	15 mins	30 mins	15 mins	20 mins	15 mins	20 mins	25
BATHROOM/TOILET	60	15 mins	15 mins	20 mins	10 mins	20 mins	15 mins	15 mins	15 mins	15

Figure 11.5 An energy audit example, of Rob's place.

Figure 11.5 shows how Rob did his calculations so he could see where to make economies to achieve the greatest gains. People still switch off 40-watt light globes (which is a good thing to do) believing they are saving substantial energy, whereas they would achieve greater economy and comfort by investing in insulation and their money would be returned within two years.

Storing heat

After choosing the most efficient and least polluting heating system, you want to save the heat you have generated. This can be achieved by installing insulation and draught-proofing your home.

- Insulating your *ceiling* is the best method of retaining heat and conserving energy. (The best roof insulation is a sod roof with 10 centimetres of dirt, but sod is not very practical in a suburban home.) Early insulating materials are suspected of emitting toxic gases or dangerous particles, such as glass particles and polystyrene fibres, so choose insulating materials that are as near as possible to natural materials and, if possible, are by-products of other industries; for example, cellulose, coconut fibre and wool.

- *Floors* can be insulated with carpet and underlay or from below with ceiling insulation held in with chicken wire. Cement slabs need to be insulated 60 centimetres inwards from the outside edge.

- *Walls* are naturally insulating if made of stone, mud brick, straw bale or adobe. These materials absorb heat and radiate it back later.

- *Window edges* should be lined to prevent winter draughts. Heavy lined curtains with pelmets assist considerably with keeping the warmth in. Double glazing is most effective on south-facing (north-facing in the northern hemisphere) windows. Other places to draught-proof are doors and the edges of ceilings.

TABLE 11.2: USING GADGETS

Item	Risk	Change to
Colour TV	*	Sit more than 2 metres from screen
Black and white TV		Go to the theatre
Video	*	Play-reading at home
Microwave oven	*	Open door away from you; cook outside with family
Computer	*	Take frequent breaks
Radio		Sing or play instruments
Electric blankets	*	Preheat bed, switch off at wall; hot water bottle; flannel sheets; one dog
Hair dryer	*	Use towel or sun
Clothes dryer	**	Use the sun; air clothes
Power tools	**	Change to hand tools and get fit
Vacuum cleaner	**	Straw broom

Cooking

Cooking is expensive and air polluting. Gases can build up to unhealthy levels in tightly closed kitchens, so ventilation is a priority for most kitchens. Alternatively, outdoor cooking is pleasant and desirable. Remember that takeaway food outlets contribute substantially to air pollution compared with home cooking.

In recent years several appropriate technology centres have designed very efficient wood-heater cooking and water-heating units. Efficient wood-fired stoves are conscionable if you plant trees to replace those you burn; that is, if you design and plant a woodlot. If you are not already planting so that future generations will have the equivalent tree resources then it is time to start. Any fossil fuel is environmentally expensive with coal and coke costing more than the others.

Maximum energy efficiency is achieved by using one energy source to serve several functions. For example, I use a closed combustion stove to heat my water for washing, washing up and cooking, and of course it heats my house in winter. The incongruous part is that my electricity bills go down in winter when I use the stove and go up in summer when the stove is not alight.

Using gadgets

There is clear evidence that low-level radiation leaking from electrical items in the home is damaging to health. In particular, as with chemicals, children are most at risk because they are growing and their bodies absorb more toxins and radiation than adults. Table 11.2 is a list of risks associated with common household goods. For the sake of your health, the fewer you have the better and, in particular, those marked with an asterisk are believed to leak or be more damaging than the others. Those with two asterisks draw large amounts of electricity.

Materials

These are all the food, medicines, cosmetics, clothes, toys and packaging we bring into our homes. Before you buy them ask yourself the following questions:

- Is it biodegradable? If not would I be happy to bury it in my backyard?
- Can it be recycled when I've finished using it?
- Am I happy to eat it, drink it or have it close to my skin?
- How long will it last and what repairs will it need?
- Can I repair it, reuse it or recycle it?

If the answers are 'no', then do not buy it.

TABLE 11.3: HOUSE CHECKLIST

HOUSE PROBLEM	WHAT TO DO	FIX: WHEN AND WHO AND WHERE
Too hot in summer	Insulate ceiling Pergola and vines Deciduous trees Shady verandahs Breezeways/corridors Solar fans Cool air tunnels Deciduous creepers Shade house with plants	
Too cold in winter	Insulation Double glazing on coolest side Thermal mass in doors/walls Lined curtains and pelmets 70% glass wall on sunny side Compact house Gap sealing Glasshouse to sunny side	
Too humid (fungi, moulds)	Cross-ventilation Solar fans Exhaust fans Fix window open Let more sunlight in	
Too dry	Indoor plants Indoor ponds Duct air from glasshouse Dense planting	
Artificial light during day	Work in better-lit room Skylights Paint walls a light colour Light furnishings Replace verandah with pergola Insert wall or glass door	
Severe winds	Build mounds Plant windbreaks Insulation Trellises Suntraps	
Pollutants: Exterior Interior	Dense fine-leaf plants Change to natural materials Improve ventilation Reduce electric gadgets Cook own food Prepare own cosmetics and medicines	
Other house problems: Noise	Fences Insulation Double glazing Complain to responsible body	

HOUSE PROBLEM	WHAT TO DO	FIX: WHEN AND WHO AND WHERE
Poor work/leisure areas	Retrofit Change use of area	
Sensory deprivation (smell, sound and sight)	Put in windows and doors Skylights, pergolas Attach glasshouse	
Chemicals: Furnishings Building materials Paints	Change to natural fibres and dyes Mud brick, adobe, stone, wood Whitewash and tints	
Cleaning agents	Natural glues, waxes, sealants Pure soap Borax Methylated spirits Bicarbonate of soda Vinegar	
Machines: Photocopier Computer	Provide excellent air circulation Frequent rests	

Ideally, we would never bring home anything we can't dispose of within our own boundaries. In reality, this is difficult; for example, what do you do with old toothbrushes? I try to replace as much plastic as I can with biodegradable materials. I use glass containers, cotton carrybags, belong to a food co-op and take all my own containers. I buy nail- and toothbrushes with natural fibres. Since I found that tins are very slow to break down buried in holes in the ground I resist most tinned food and, of course, anything packaged in plastic. It means that my shopping is smaller, lighter and cheaper. I put rubbish out about three or four times a year. My medicines and cosmetics are few and simple, and my clothes are of natural fibres and can be mulched. And I like organic chocolate.

Transport

This is the worrying one because it is the really big monster resource junkie. Everyone feels their own car or motorcycle is absolutely indispensable. If your home is quite energy hungry it will cost you about 14,000 kilowatt hours per year in a cool climate. A two-car family will consume 40,000

kilowatt hours per year. However, it is very easy to make considerable savings for your health and the environment by walking, cycling, or taking a bus, train or taxi instead of your own car. It is also often more leisurely and enjoyable. Try it and see.

 Try these:

1. Make a list in your journal of ways you can reduce household waste. Count how many times you put out rubbish in a month and then reduce it.
2. Design a new home for your family which is clean passive solar, and will cost almost nothing in maintenance in the future. Now furnish it.
3. Take your place, or another, and retrofit it to become a sustainable clean green home.
4. Do your own energy audit, like Rob's (see Figure 11.5).
5. Reduce your car use and start with one day a week 'car free'.
6. Use Table 11.3 to shift to a more Earth-friendly household. In the third column, mark items by your priorities and write the dates the work will be started and completed.

CHAPTER 12

Your garden: Zone I

When I think about the food I need for the day my mind goes to what is growing in the garden, and not to what is stashed in the refrigerator. So I visualise my garden as a food shop. In fact, my refrigerator is rarely switched on and it is also fairly empty. Most people in the world living in hot climates do not have refrigeration and manage very well with their food gardens and local markets selling fresh food.

We have three main types of agriculture in the world: agribusiness, the home lawn and the house garden.

Agribusiness is only concerned with food as a commodity in the marketplace. It relies on high inputs of chemicals and is energy hungry. Of its income, 30–40 per cent goes to multinationals. Its greatest expense is servicing debt in the form of interest payments because, worldwide, almost every commodity farmer is in debt. It is only viable with government subsidies. It is the greatest threat to Earth because it causes excessive land degradation.

The home lawn consumes enormous chemical, water and energy inputs yet answers no food need at all. The chemicals such as fertilisers, fungicides and weedicides used per hectare are double and sometimes treble those used in intensive agriculture. Lawns are monsters with an insatiable appetite. In California, 14 per cent of first-class water is used on lawns. Until it started running out, Western Australia poured 20 per cent of its drinking water on lawns. Lawns cost a minimum of $6.00 per square foot to maintain.

The home garden is what permaculture has, in many people's minds, come to represent. Although it isn't all of permaculture, it is an important part. The home vegetable garden yields at least $5.00 per square foot and is the only form of agriculture which directly feeds people. You can grow 80 per cent of your food within 50 metres of your back door. In 1998 the estimated yield from home gardens was $92 million.

Our ethical task is to:
- grow as much food as we can at home
- ensure that home gardens carry out several vital ecological functions.

Our design aims for a vegetable garden are for it to:
- occupy a permanent position with self-seeding biennial and perennial plants
- be non-polluting for plants, water or soils
- be abundantly productive
- produce surplus to give or sell locally
- reduce food miles and our ecological footprint
- recycle all household organic matter.

If we don't have design aims for a vegetable garden:
- we can waste money and resources
- we can become dependent on polluted, industrial food.

CHARACTERISTICS	RESULTS
SMALL AND INTENSIVELY CULTIVATED FOOD GARDEN (VEGETABLES, HERBS AND SMALL FRUITS).	• INCREASES SELF-RELIANCE. • MAKES USE OF HOUSEHOLD ORGANIC WASTES. • PROVIDES HIGH YIELDS PER UNIT AREA.
A BASIC STRUCTURE OF PERENNIAL, BIENNIAL, SELF-SEEDING AND SELF-MULCHING PLANTS.	• REDUCES HUMAN LABOUR. • INCREASES ENVIRONMENTAL STABILITY.
ABUNDANT AND DIVERSE PLANTINGS.	• ALLOWS NATURAL PROCESSES TO SELECT THE PLANTS MOST SUITED TO THE SITE.
NOT MORE THAN FIFTY METRES FROM THE HOUSE.	• GARDEN BEDS ARE NOT EASILY OVERLOOKED. • EASY TO DIRECT GREY WATER TO GARDEN BEDS; HARVEST PRODUCE; WEED PLANTS; AND PROTECT PLANTS AND ANIMALS FROM WEATHER EXTREMES AND PREDATORS.
VISITED FREQUENTLY.	• VEGETABLES AND FRUIT CAN BE HARVESTED AS REQUIRED.
CONNECTED BY CIRCULAR, WINDING OR SPIRAL MULTI-PURPOSE PATHS.	• COMPOST BINS, GARDEN BEDS, POULTRY YARD, FISHPOND, ETC. CAN BE VISITED IN ONE WALK.
MAINTENANCE AND CLEARING (TRACTORING) DONE BY ANIMALS.	• WEEDS, INSECT PESTS AND DISEASED PLANTS ARE CLEANED UP BY ANIMALS, RATHER THAN PEOPLE. • ANIMALS PROVIDE ADDITIONAL YIELDS OF EGGS, MEAT, MANURE, ETC.
SHEET-MULCHED GARDEN BEDS.	• REDUCES WATERING AND WEEDING. • PROTECTS SOIL FROM EROSION AND LOSS OF VALUABLE NUTRIENTS.

Figure 12.1 Characteristics of Zone I.

Ecological functions of a home garden

A home kitchen or vegetable garden is fundamental to your design because, as well as providing you with fresh vegetables, it carries out other important functions. It:

- offers you the security of quality, quantity and supply of chemical-free food
- transforms your waste organic materials into mulch and humus
- builds self-reliance and creative leisure
- absorbs grey water and turns it into biomass (living organisms)
- cleanses grey water before returning it to waterways and the water table
- releases you from the bondage of lawnmowers, edgers and suchlike with their

smell, noise, fuel consumption, expense and possible danger

● provides habitat and niches for wildlife and insect predators

● conserves biodiversity and heirloom varieties

● reduces the stress on marginal land used for growing food because cities and suburbs increasingly take the best land

● lowers the overall burden of environmental damage associated with growing food as agribusiness.

The kitchen garden has all elements working in productive, satisfying and efficient ways and will demonstrate the following characteristics:

PERMANENCE It moves towards a balanced state where the garden perpetuates itself, thus ideally dispensing with the need for a gardener at all. This is achieved by plants which are self-seeding, and biennials and perennials.

ABUNDANCE Abundance is achieved through dense and diverse planting which acts as a buffer in adverse conditions and yields under all conditions. Don't aim for large harvests of one particular fruit or vegetable, although you will certainly get this sometimes. Harvest a large amount of many species and their parts.

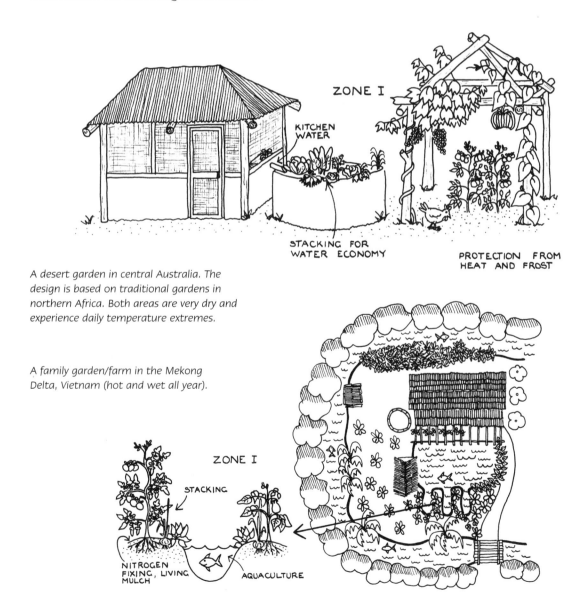

A desert garden in central Australia. The design is based on traditional gardens in northern Africa. Both areas are very dry and experience daily temperature extremes.

A family garden/farm in the Mekong Delta, Vietnam (hot and wet all year).

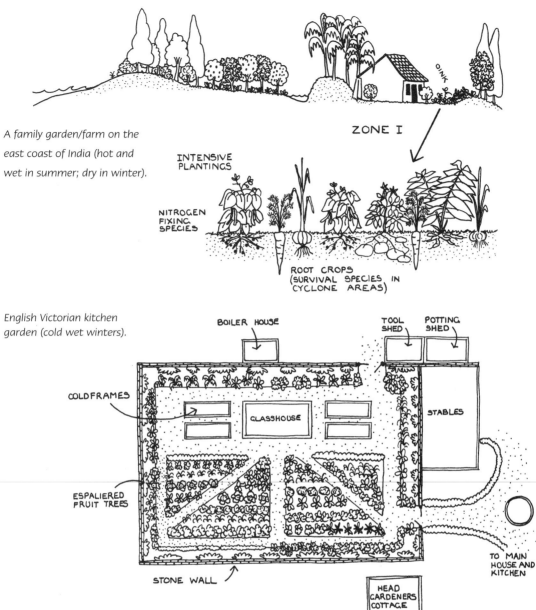

A family garden/farm on the east coast of India (hot and wet in summer; dry in winter).

ZONE I

INTENSIVE PLANTINGS

NITROGEN FIXING SPECIES

ROOT CROPS (SURVIVAL SPECIES IN CYCLONE AREAS)

English Victorian kitchen garden (cold wet winters).

BOILER HOUSE

TOOL SHED   POTTING SHED

COLDFRAMES

GLASSHOUSE

STABLES

ESPALIERED FRUIT TREES

TO MAIN HOUSE AND KITCHEN

STONE WALL

HEAD GARDENERS COTTAGE

Figure 12.2 Four types of traditional vegetable gardens.

EVERYTHING GARDENS This means plants and animals are carrying out functions usually seen to be the work of humans. Animals cultivate the earth with their feet, beaks and burrows, and plants use their roots and associated micro-organisms such as fungi. Plants and animals nourish each other and the soil with products. They prune and harvest fruit, leaves, seeds and limbs and propagateplants by carrying seeds, spores, runners, layering and eating.

Over the centuries, wherever people have gardened sustainably they have employed the same principles, regardless of climate or culture. You can see this in the four types of gardens illustrated in Figure 12.2. Unlike modern agribusiness, these gardens have sustained people without degrading the land:

- desert garden in Australia, based on traditional oasis gardens of northern Africa—arid climate with extremes of temperature
- Mekong Delta of Vietnam—conditions are hot and wet all year round
- monsoonal east coast of India—hot, wet summers and dry winters
- English Victorian kitchen garden—a cool temperate climate with wet, cold winters and summers.

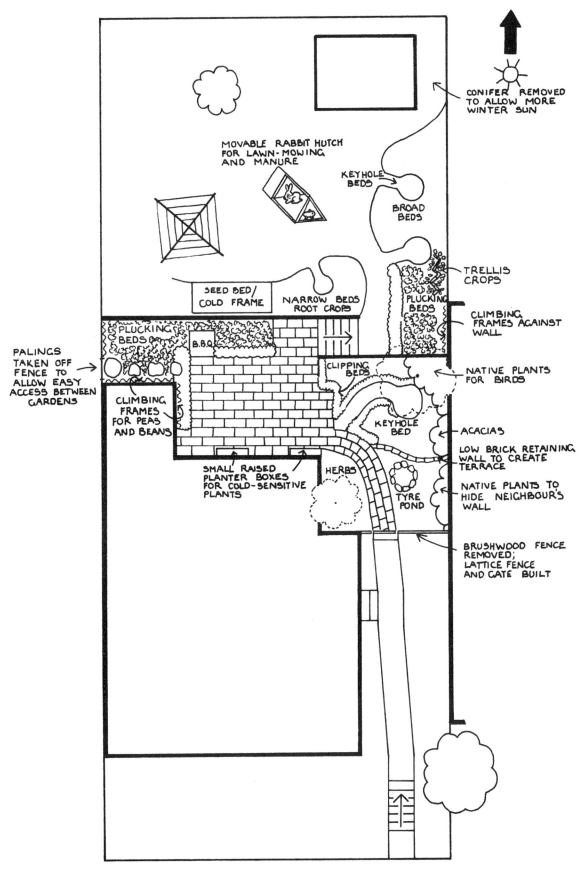

CONIFER REMOVED
TO ALLOW MORE
WINTER SUN

MOVABLE RABBIT HUTCH
FOR LAWN-MOWING
AND MANURE

KEYHOLE
BEDS

BROAD
BEDS

TRELLIS
CROPS

SEED BED/
COLD FRAME

NARROW BEDS
ROOT CROPS

PLUCKING
BEDS

CLIMBING
FRAMES AGAINST
WALL

PLUCKING
BEDS

B.B.Q.

PALINGS
TAKEN OFF
FENCE TO
ALLOW EASY
ACCESS BETWEEN
GARDENS

CLIPPING
BEDS

NATIVE PLANTS
FOR BIRDS

CLIMBING
FRAMES
FOR PEAS
AND BEANS

KEYHOLE
BED

ACACIAS

LOW BRICK RETAINING
WALL TO CREATE
TERRACE

SMALL RAISED
PLANTER BOXES
FOR COLD-SENSITIVE
PLANTS

HERBS

TYRE
POND

NATIVE PLANTS TO
HIDE NEIGHBOUR'S
WALL

BRUSHWOOD FENCE
REMOVED;
LATTICE FENCE
AND GATE BUILT

Figure 12.3 Rob's place, Zone I design.

SHEET MULCHING

WHAT TO DO	WHY DO IT	WHAT IT LOOKS LIKE
• SLASH LONG GRASS AND WEEDS, MOW LAWN AND LEAVE CLIPPINGS IN PLACE	• CLIPPINGS DECOMPOSE AND ADD ORGANIC MATTER TO THE SOIL	
• WET WHOLE AREA THOROUGHLY	• RAIN WON'T REACH THE SOIL THROUGH THE LAYERS	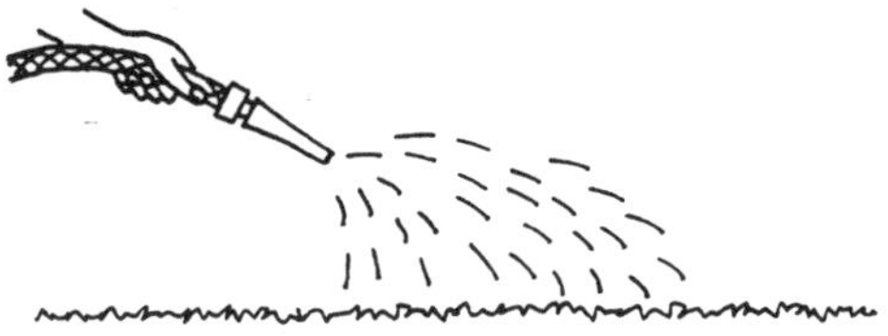
• ADD SOME AGRICULTURAL LIME	• HELPS BIND ANY HEAVY METALS SO THEY CANNOT BE TAKEN UP BY PLANTS	
• SOAK PAPER, CARDBOARD, UNDERFELT OR EVEN OLD CARPET. LAY OVERLAPPING SHEETS OVER WHOLE AREA	• STOPS WEEDS AND ADDS MORE ORGANIC MATTER TO THE SOIL	
• MARK OUT PATHS WITH LIME, STONE, BRICKS OR TIMBER	• PREVENTS BEDS BEING BUILT OVER PATHS	
• THROW ANY ORGANIC WASTE SUCH AS GRASS CLIPPINGS, GARDEN SCRAPS OR WEEDS ON GARDEN BED	• IT WILL ALSO DECOMPOSE, AND TURN INTO HUMUS	
• ADD OLD HAY OR GRASS TO 15cm DEEP	• MORE COMPOST TO TURN INTO HUMUS	
• ADD 10-15cm OF ROTTED MANURE, COMPOST OR MUSHROOM COMPOST (ALWAYS DIFFICULT TO GET ENOUGH!)	• IMMEDIATE SOURCE OF PLANT NUTRIENTS	
• ADD LAYER OF CLEAN WEED-FREE MULCH (10cm) SUCH AS STRAW, RICE HULLS, OAT HUSKS OR SUNFLOWER HUSKS	• HOLD WATER IN, RETAIN VOLATILE NUTRIENTS, PROTECT SEEDLINGS, SOIL TEMPERATURE CONTROL	

Figure 12.4 Sheet mulching.

Planning your Zone I garden layout

Review the sun and wind patterns on your property, the ease of access, your whole site water plan, soils and microclimates—and draw up a plan keeping these ideas in mind.

- *Start small* and get it right. There is nothing like success to carry you forward and to ensure that you have enough resources to complete the next stage. Mistakes or problems can also be corrected while they're small.
- Begin your plan with *permanent structures.* Think carefully and then place some or all of the following permanent structures where you want them. If you have planned a structure but don't have the time or money to build it now, then keep the space by undertaking only short-term activities there.

  - water gardens
  - greenhouse
  - shadehouse
  - worm bed
  - recycling area
  - cooking space
  - compost
  - herb spiral
  - clothesline
  - cold frames/hot beds
  - garden shed
  - water tanks
  - outdoor toilet
  - fixed animal housing
  - pergolas/trellises
  - keyhole beds.

- Design *paths* as circular, winding or spiral so they enable you to take an interesting walk to accomplish several things with one saunter around the garden. On the way to collect eggs or feed poultry, hang out the clothes and visit the cold frames.
- *Eliminate small fussy lawns* because they waste valuable space and require too much maintenance.
- *Group* similar activities and decide how activities will support each other. For example, place the potting shed near the greenhouse, the worm farm near the vegetable garden, and the compost bins right in the middle of the vegetable garden. Your outdoor cooking area can be close to the kitchen door.

If you have done a thorough site analysis you will design well. Figure 12.3 shows structures sited after a thorough site analysis.

Making your garden

Sheet mulching is the permaculture technique used in Zone I for improving and building soils (usually a long, slow process). In normal 'hard work' gardens, soil improvement is achieved by tilling or digging, then perhaps hoeing and finally raking the whole area, then leaving it bare until you wish to plant. However, we know that, except for deserts, nature never leaves the soil bare and vulnerable to damage.

Sheet mulching simply involves covering the existing ground surface, whether it is old roadway, concrete or grass, and building a new clean soil over the old base. The technique is called 'sheet mulching' because a cover sheet of mulch is laid over the garden. Figure 12.4 shows the nine steps I use in sheet mulching. Other people vary them slightly.

All the layers must be thoroughly wet as you build up your garden. However, in the long term the garden will require far less watering than a normal 'hard work' garden. And you can plant into your new sheet-mulched garden immediately.

Planning your planting

Siting vegetables, flowers, fruits and herbs depends on the following factors:

- frequency of harvesting and use
- level of maintenance
- plants' life expectancy
- growth habit (or adult shape)
- space required when mature
- plants' requirements for water, sun and wind.

The following diagrams illustrate a 'model' Zone I food garden. You can use all or just some of these ideas in your own garden.

- *At the kitchen door:* Plant a citrus tree such as a lime or lemon with small herbs such as chives and parsley underneath. A variety

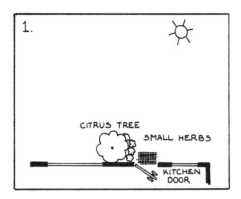

specially chosen for your area will crop 2–3 times a year. Citrus are best stored on the tree. Harvest them frequently because they have high vitamin C content.

Figure 12.5
Herb spiral.

- *Culinary and medicinal herb spiral:* These herbs, which you use daily for health or cooking, are planted in a spiral on the other side of the kitchen door. The spiral has many aspects and niches and allows for a variety of microclimates from very hot on the west to dry at the top. It stacks plants vertically as well. Herbs grown here include all the cultivars of marjoram, oregano, rosemary, sages, basils, savouries, thymes and tarragons. Inter-plant annual and perennial herbs.

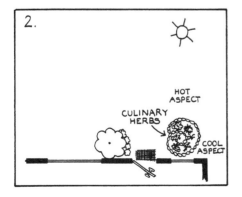

- *Clipping beds:* Position these on the edges of the paths and inside keyhole beds. They are mainly perennials and clipped for their edible leaves. They require worm castings, potash and lime about twice a year. Planted next to paths they receive lots of sun, are highly accessible for frequent clipping and are usually protected from wind. Suitable plants are chives, sorrel, corn salad, dandelion, salad burnet, mustard greens and nasturtiums.

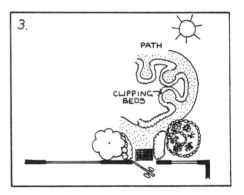

- *Plucking beds:* These are placed just behind the clipping beds, still close to the house, and consist of fast-growing and taller plants that are frequently harvested without pulling the whole plant. You can pluck leaves, seeds and fruit. Plant broccoli, silverbeet, Swiss chard, rose chard, kale, English spinach, Brussels sprouts, bunching onions, celery, non-hearting lettuce, coriander and zucchini, which all need frequent harvesting or they grow too big.

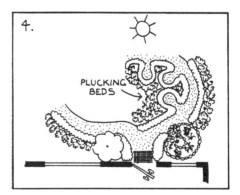

- *Narrow beds:* Narrow beds are for plants that grow vertically or have high light requirements. The beds are aligned north–south to receive both morning and afternoon sun. It is valuable to have some permanent plants here such as asparagus, which has a lifetime of about 20 years. You can plant beans, peas, carrots, tomatoes, radishes, climbing peas and beans, asparagus, okra and eggplant.

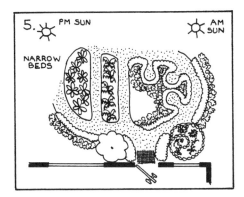

- *Broad beds:* These are for vegetables that take a longer time to ripen and are only harvested once. They are slower growing and do not need too much attention. Some of these plants are hearted lettuces, cabbages, lupins, sweet corn, pumpkin, sugar cane (eaten raw), Chinese cabbage and cauliflower. You can add Jerusalem and globe artichokes and use them as low suntraps and temporary windbreaks.

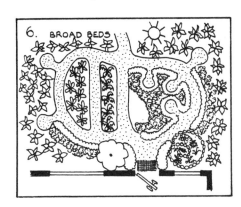

- *Broad-scale beds (staples):* These beds are possible if you have half a hectare or more land available. Growing grains moves you closer to self-sufficiency. Staples can be grown either by alley cropping (see page 152) or by the Fukuoka methods (see page 155). Choose the most successful and best adapted

for your area from corn, wheat, rice, oats, barley, millet, sorghum, potatoes and sweet potato. Corn is the highest yielding.

- *Vertical growing/trellis crops:* Fences, pergolas and sides of buildings increase the growing area and take advantage of a plant's needs for a special micro-climate. They assist garden productivity by effectively increasing the size of the garden. Crops you can grow are climbing peas and beans, passionfruit, choko (chayote), brambles, kiwifruit, jicama, New Zealand spinach, cucumbers, pumpkin and grapes.

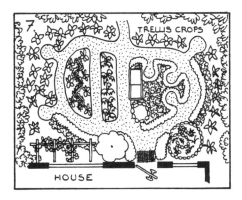

Now look again at Figure 12.3 and see how Rob's Zone I compares with the 'model' Zone I garden described above. Rob modified the model because of limiting factors which he found in his site analysis.

Permaculture gardening hints

Here are a few tips and hints to keep in mind as you establish your home garden.

Crop rotation

This means changing the place where you grow plants. Crops are rotated according to their families, nutrient needs and build-up of pests. For example, for garden hygiene, change the place where potatoes grow each year so pests do not build up. In general the order of rotation is: legumes, cabbages, tomatoes, onions, root vegetables, and then start again with legumes.

Grey water

Grey water from the house is harmless only if you use mild pure vegetable soaps for all washing. Water your Zone I garden only when the soil is dry down to the second joint of the forefinger. This encourages roots to search deeply for soil water and they will be more drought-resistant.

Weed management

This is best performed by repetitive sheet mulching, dense planting, and judicious use of small animals such as caged rabbits, quail and guineapigs. Move the cages as necessary.

Companion plants

These are plants that help other plants in one or more of the following ways:

- the smell of their volatile oils discourages pests
- they are nitrogen-fixing plants of the legume family and supply nitrogen to other plants
- they have shapes that confuse pests' recognition ability.

The benefits of companion planting are achieved by inter-planting using some of the well-known herbs, and adding herbs and flowers to your vegetable garden for their multiple benefits.

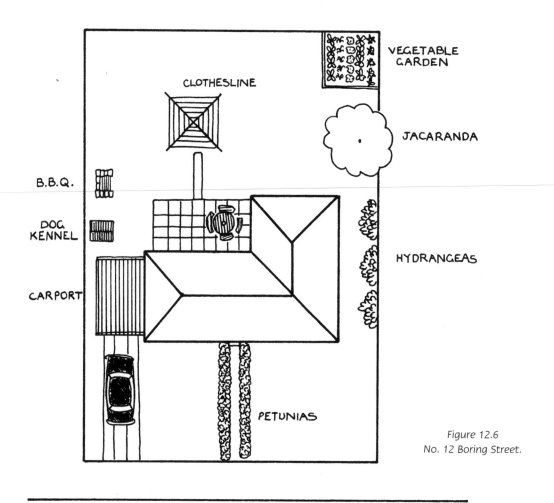

Figure 12.6
No. 12 Boring Street.

BORING STREET

Indigenous plants

Native plants are fundamental to every garden since they provide habitat and food for wildlife threatened by loss of habitat. They can be planted as a food garden hedge to supply food for both animals and humans. Most importantly, they assist in maintaining biodiversity on a regional scale. Become a propagator of local species.

Fruit

You don't need a large garden to grow fruit. Berries don't take up much room and they are ideal on trellises. Depending on your microclimate, choose from Cape gooseberry, brambles, English gooseberry, currants, strawberries or whatever grows well near your place. Most fruit trees are available as dwarf stock suitable for growing in pots or you can plant fruit trees with multi-grafts. Fruit trees can also be espaliered against walls or grown in hot houses.

Try these:

1. Look at Figure 12.6, an average house and garden block. In your notebook turn it into a permaculture food garden by redesigning it.

2. Now redesign your food garden, taking into account the limitations you found in your site analysis and seeing whether you can turn them into possibilities. Follow as closely as you can the garden design steps given in this chapter.

3. Write down all the vegetables, herbs and flowers you would like to harvest from your garden. Then make a growing calendar according to harvest season to show how you can keep yourself in food all year round.

CHAPTER 13

The food forest: Zone II

All plant and animal species have evolved in specialised ecosystems, often forests and woodlands, with other plants and animals complementing their functions and serving each species' needs. When we take a plant out of its natural ecosystem, we remove it from this support system.

In most cases the plant will only survive if we do all the work. In permaculture we design an orchard as a sustainable food forest with a wide diversity of plants and animals supporting and complementing the needs of other species. As its productivity increases, work inputs and maintenance decrease. The food forest is called Zone II because it comes after Zone I in the use of work and resources. A food forest is:

- a *waru*, or guild, of interrelated and interdependent fixed and mobile species which work for the trees just as the trees work for them
- a sustainable system of productive permanent trees which provides excellent return for effort
- more successful when there is thoughtful observation and use of local knowledge
- less susceptible to losses and failures from pests and diseases than a monoculture orchard.

 Our ethical task is to:
- plant productive, high-yielding fruit trees for our own use and for the future to return what we have used
- to reduce 'food miles' and unfair trade.

 Our design aims for a food forest are to:
- grow a diverse range of food trees
- create environmental stability, greatly reducing the incidence of pests and the need for artificial chemicals
- plant living groundcover mulches to increase the fertility of the soil and the vigour of the trees
- observe and analyse local conditions and microclimates to assist in selecting and placing appropriate species to reduce crop losses and tree failure.

 If we don't have design aims for a food forest:
- we neglect future generations' need for fruits and nuts
- we contribute to climate and environmental instability
- we can lose valuable heritage species forever
- we encourage monopolies, monocultures and trade of fruit as a commodity not a resource.

CHARACTERISTICS	RESULTS
MANY SPECIES AND CULTIVARS OF FOOD AND NON-FOOD PLANTS.	• PROVIDE FRUIT, ANIMAL FODDER, MULCH, WIND PROTECTION AND HABITATS FOR BENEFICIAL PREDATORS.
SMALL ANIMALS INCLUDING BEES AND FREE-RANGING POULTRY.	• PROVIDE FERTILISER FOR PLANTS. • CONTROL PESTS AND WEEDS. • PROVIDE EDIBLE ANIMAL PRODUCTS. EG. MEAT, HONEY AND EGGS. • AID POLLINATION AND SEED DISPERSAL.
HEAVILY MULCHED (LIVING GROUNDCOVERS, TREE MULCHES AND FLORAL PASTURES).	• PROTECT SOIL AND REDUCE NEED FOR WATERING. • PROVIDE HABITAT FOR PEST PREDATORS. • PROVIDE FORAGE FOR FREE-RANGE POULTRY. • FLORAL PASTURE PROVIDES BEE FODDER. • LEGUME GROUNDCOVERS FIX NITROGEN IN THE SOIL.
GRAFTED, LOCALLY PROVEN AND HEIRLOOM PLANT VARIETIES.	• GRAFTED PLANTS PROVIDE MORE CONSISTENT YIELDS. • LOCALLY PROVEN PLANTS ARE HIGHLY ADAPTED TO THE LOCAL ENVIRONMENT. • HEIRLOOM VARIETIES NEED TO BE PRESERVED TO MAINTAIN THE GENE POOL.
MULTI-PURPOSE WALKS.	• ALLOW SEVERAL ACTIVITIES TO BE COMPLETED IN THE ONE WALK (EG. COLLECT EGGS, FRUIT, SEED AND MULCH, AND OBSERVE AND REFLECT ON WHAT HAS BEEN DONE).
STACKING OF PLANTS (STORIES OF GROUNDCOVERS, SHRUBS, CREEPERS AND TREES).	• ALLOWS INTENSIVE USE OF AVAILABLE SPACE. • INCREASES PLANT PRODUCTIVITY (EACH PLANT CAN UTILISE THE SURROUNDING RESOURCES - WATER, LIGHT, NUTRIENTS, ETC.- TO ITS FULL POTENTIAL).

Figure 13.1 Characteristics of Zone II.

Siting Zone II

The food forest is placed according to the same criteria used for establishing Zone I: by considering the site's potential productivity, water and energy requirements, and maintenance inputs.

It will not require as many inputs or work as your food garden so it is placed just beyond Zone I and before Zone III. Figure 13.2 shows the ideal land profile for siting the orchard in relation to windbreaks, water (dams, swales), aspect, slope and access.

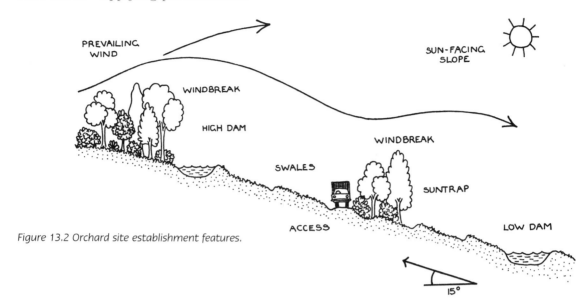

Figure 13.2 Orchard site establishment features.

Slope

Gentle slopes are ideal for planting because they have a range of drainage conditions (such as dry at the top of the slope and increasingly moist towards the bottom), soil types and microclimates.

Water

Water harvesting, through dams and swales for water retention, are placed on the high points of the property. Water can then be fed to the orchard by gravity.

Land preparation

It can take two to three years on degraded land to prepare the soil, water systems and shelter for the young trees.

Windbreaks

Hardy windbreak trees are planted to lift and deflect prevailing hot and cold winds, which can destroy the whole crop at blossoming and fruiting time. Mixed species in the windbreak provide a variety of yields and functions, such as honey, habitat, fuel and

Figure 13.3 The food forest: a cultivated waru.

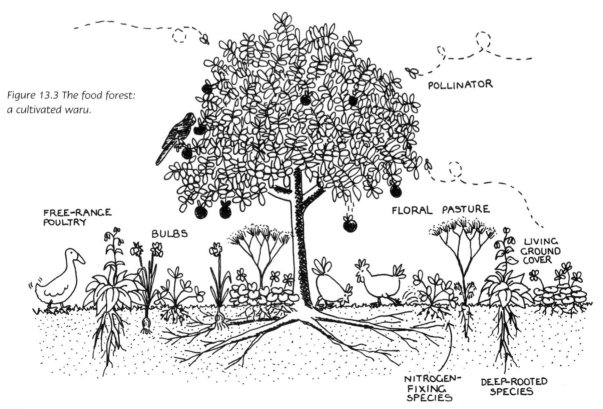

grazing mulch. In-crop windbreaks of nitrogen-fixing trees supply nutrients and protection. Windbreak planting patterns are parabolic so that the wind is diverted around the plantings (see figures 13.4 and 13.5). Parabolic-shaped windbreaks also act as a suntrap for the fruit trees.

Aspect

Most fruit trees require maximum sunlight and are planted on sun-facing slopes.

Soils

Most agricultural soils are depleted in nutrients and are compacted. Ripping with an agroplough and planting with green manure or cover crops will help overcome these problems. A crop of densely planted potatoes is particularly useful for loosening compacted soil. Zone II soils are protected and improved by providing mulches. However, the area is normally too large for sheet mulches so spot mulches and living mulches are used.

Mulches

In orchards living groundcover mulches are planted under the orchard trees to provide:

- all-season food for bees which will pollinate the fruit trees
- food for poultry
- companion plants for the fruit trees
- nitrogen for the soil.

A. PROFILE

*Figure 13.4
Orchard windbreaks.*

Profile: As well as giving shelter from the prevailing winds, the trees in the windbreak can provide other benefits. In this diagram, the tree lucerne (middle plant) will fix nitrogen in the soil, and the mulberry tree (right) and the blueberry bush (left) will provide forage for poultry.

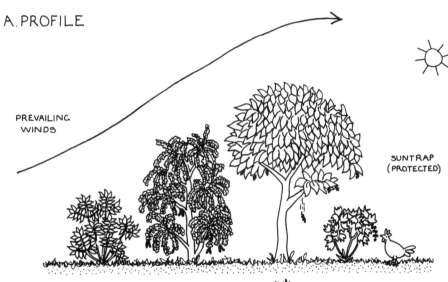

PREVAILING WINDS

SUNTRAP (PROTECTED)

B. PLAN VIEW

Plan view: The windbreak is planted in the shape of a parabola to deflect wind sideways around the orchard.

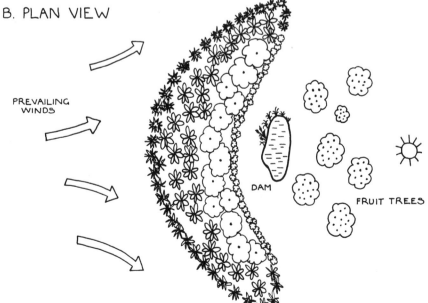

PREVAILING WINDS

DAM

FRUIT TREES

PREVAILING WIND

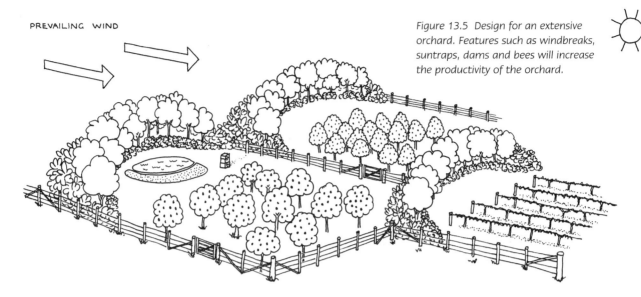

Figure 13.5 Design for an extensive orchard. Features such as windbreaks, suntraps, dams and bees will increase the productivity of the orchard.

Suitable plants include comfrey, bulbs (jonquils, hyacinths, irises, daffodils, babianas, freesias), sages, thymes, onions, chives, and self-seeding species such as fennel, dill, borage, carrots, nasturtiums, balm clovers, legumes and daisies.

Tree selection and placement

All trees have needs that are usually served or organised by farmers or gardeners. However, each species or cultivar needs different ways or amounts of meeting these needs. This is why some research is required before choosing the fruit trees for your orchard. Some of the needs include:

- pruning
- pollination
- nutrients
- pest management
- light breezes
- drainage
- protection.
- seed disposal
- harvesting
- water
- sunlight
- space
- soil micro-organisms

It is best to start with hardy, locally proven species known to grow well in your area. Although their fruiting characteristics may not be as desirable as other varieties', they will have a greater chance of establishing in the new environment. Later you can add varieties that have special qualities, such as early or late fruit, special flavour or colour, and good storage or processing characteristics. Finally,

in later years, when the soil and microclimate have been modified by the trees you have planted, you can add more peripheral species to take advantage of such factors as non-average seasons or atmospheric warming.

If possible, choose grafted species because they bear more heavily and the rootstock is hardier.

If you know the species' geographical origin you may be able to modify your microclimate to suit it. For example, although almonds prefer a mild Mediterranean climate they can still be grown in cool areas by planting near a warm, west-facing wall with good drainage and excellent wind protection.

Select cultivars that are known to be highly resistant to diseases and pests. This will reduce the need for sprays. For example, if you grow grapes in an area with hot, wet summers you can expect the plants to suffer from fungal diseases every year. Instead, you could try to find a cultivar which has a known resistance to fungal diseases and plant it in a place where it receives drying breezes.

Some fruits, like citrus, plums and peaches, have been bred to thrive over a wide range of soils and climates. Other trees are more specialised in their requirements; for example, some grow best at high altitudes, coconuts will only fruit in warm coastal regions.

Figure 13.6 shows the climatic origins of a range of cultivated fruit trees with those requiring well-drained to dry soil placed at the top of the slope, and trees requiring cool wet conditions placed at the bottom of the slope.

By making use of the slope in this way, trees from one climate group can often be grown in the climate listed alongside in Figure 13.6. For example, many desert (hot, dry) species can be grown in Mediterranean climates. (Note that each species may have hundreds of cultivars and varieties.) After you have decided on the best positions for the trees according to the slope characteristics—warmer, cooler, deeper soil, etc.—plant according to the following traits:

- *Leaf drop*: The first deciduous trees to drop their leaves are placed in front (towards the sun) of those that drop their leaves later in the season, or which are evergreen.
- *Adult size and shape*: Small trees are planted in front of larger trees so they are not blocked from the sun (see Figure.13.7).
- *Fruit ripening*: Trees where fruit ripens

outside the leaf canopy, such as oranges, need more sun than trees with fruit ripening inside the leaf canopy.

Planting fruit trees

The general rule is to plant deciduous trees in winter and evergreen trees in summer. Plant with the opening seasonal rains—in Mediterranean climates these are the opening autumn rains. In warm, wet climates plant trees in the cooler season.

Dig the planting hole twice as wide and deep as the container and place some compost or rotted manure in the hole. If you put compost only on the surface the tree's roots will not penetrate deeply looking for food and water. Fill the hole with water and let it drain twice. Lift the plant out of the pot, place it in the hole and then backfill the hole with soil but never above the graft. Then slowly fill the hole with water until there are no more bubbles.

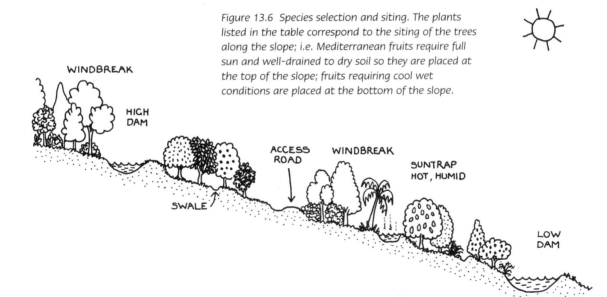

Figure 13.6 *Species selection and siting. The plants listed in the table correspond to the siting of the trees along the slope; i.e. Mediterranean fruits require full sun and well-drained to dry soil so they are placed at the top of the slope; fruits requiring cool wet conditions are placed at the bottom of the slope.*

MEDITERRANEAN	HOT DRY	HOT WET	COOL WET
OLIVES	MELONS	ROSE APPLE	APPLES
GRAPES	APRICOTS	BANANA	CHERRIES
MULBERRIES	FIGS	PINEAPPLE	PEARS
ALMONDS	DATES	MONSTERA	QUINCES
CAPE GOOSEBERRY	PUMPKINS	MANGOSTEEN	BRAMBLES
CAROBS		MANGO	BERRIES

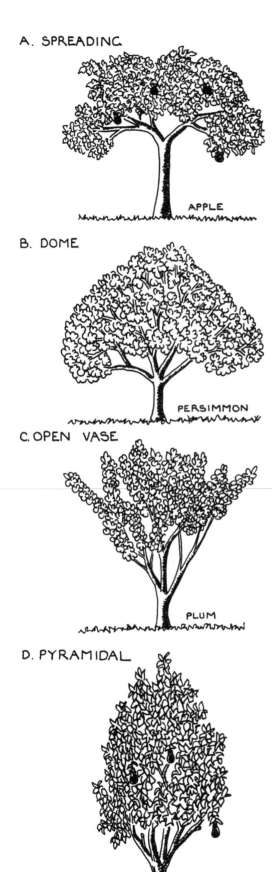

A. SPREADING

APPLE

B. DOME

PERSIMMON

C. OPEN VASE

PLUM

D. PYRAMIDAL

PEAR

Feeding the orchard

After the groundcovers and the in-crop leguminous species are well established, let your poultry into the orchard (or you may like to keep a pig, which will also help to maintain the orchard). These plants and animals will supply all the fertiliser your trees need.

You will still need to monitor the health of the orchard, however. If the land is invaded by bracken fern then the nitrogen levels are too low. You can either grow a green manure crop and chop it into the soil, or enclose your animals so their stocking rate is higher and manure output is increased. If you don't have enough animals to keep the groundcover well controlled, regular slashing of the groundcover will provide organic mulch to enrich the soil.

With a small orchard you can build compost bins or pile mulch around the trees' root zones. This saves lots of work moving compost. When the compost is ready, tip up the bin and move it to another tree.

Pollination

Pollinators are agents that carry pollen from one tree to another; for example, wind, water, wasps, birds and bees. Most insect pollinators are encouraged by flowering plants. Some, like bees, don't like and won't work in windy weather, so there is a real need for windbreaks. To encourage pollinators, flowers are stacked into orchards. Nasturtiums, daffodils, jonquils, hyacinth and irises flower early and grow well in association with fruit. Herbs such as lemon balm, fennel and dill are also food for pollinators. Bees are agents for cross-pollination of such fruits as apples, almonds and

Figure 13.7 Fruit tree habit. When you are planning the orchard you will need to consider how the shape of the mature tree will affect other trees planted nearby. For example, pear trees have a fairly dense pyramidal canopy which may cast heavy shadows on smaller growing plants. In comparison, plums have an open canopy which allows sunlight to filter through the leaves and branches.

The north-facing aspect of the author's house has been modified to increase solar gain. A brick terrace reflects heat and light into the house and acts as thermal mass, and the large sliding doors allow direct sunlight inside. A glasshouse grows food in the cooler months and warms the bathroom.

Construction of a small dam on the north-facing slope in front of the author's house. Logs from radiata pines felled on the property have been used to stabilise the steep banks, and straw mulch has been spread over the dam wall to prevent erosion and silting in the stream below.

Several smaller ponds constructed in the overflow channel below the dam are designed to slow down storm surges and trap silt. Eventually, emergent reeds and water plants will stabilise the ponds, biologically filtering water as it leaves the property.

A large-scale reed-bed system constructed to biologically filter nutrient and physically trap sediments in run-off from suburban streets before it flows into sensitive bushland in Katoomba, New South Wales (Blue Mountains City Council).

'Paying respects to the life-giving spirit of the spring'.
In Indonesia, as in many subsistence cultures, people have always been aware
of how dependent they are on springs, streams, monsoon rains and wet seasons for
their water supplies and treat them with reverence (courtesy of Impact Postcards).

Growing multiple crops in the same rows, in Vietnam.
A groundcover of legumes grows under corn on the intensively
cultivated Red River delta outside Hanoi.

Swales are ditches dug along the contour of the land to slow down run-off and allow water to infiltrate the soil (courtesy of Nada and Tony Smark, Woodbrook Farm, Harcourt, Victoria).

A tagasaste hedge provides a windbreak and protection from kangaroos for this vegetable garden (courtesy of Adrian Thomas and Marylin Tulloch, Glen Lyon, Victoria).

A retaining wall constructed from bricks.
The warmth trapped in the bricks helps to
ripen the strawberries.

A beehive in the author's backyard.
As well as yielding honey, the bees pollinate
the fruit trees and vegetables.

Above and right: A chicken yard in the author's
backyard is divided into several small enclosures in a
Zone II orchard and poultry system. Each season the
chickens are moved into a different section and the
previous one is used for growing vegetables.

The mandala garden in illustration. A banana circle grows in the centre of this garden surrounded by a large variety of vegetables. Perennial food and medicinal plants are planted around the outside and a living fence protects the garden from farm animals.

In this circle garden in Cambodia, corn and cucumbers are planted around a small pit filled with compost materials and cow manure. The compost rapidly decomposes to provide nutrients and stays moist, reducing the need to water the garden.

Varieties of pumpkin and squash. By continuing to grow and save the seeds of traditional, heirloom and rare varieties of food crops, we safeguard our choice for characteristics in our crops such as early or late maturing, insect and disease resistance, taste and culinary use.

This house in the high-density Sydney suburb of Manly has been retrofitted with solar power, solar hot water, composting toilet and a complete grey-water recycling system. The backyard has been planted with fruit trees with vegetables growing underneath.

A small pond will help to modify the microclimate, reflect light, attract wildlife, serve as a barrier to fire and a back-up water supply for the garden and animals.

Constructing a sheet mulch garden is simple and quick, and can be a social event, as shown here at Rob's place.

A Zone I vegetable garden with a diverse Zone II orchard in the background (courtesy of the Permaculture Community Garden at the Ecoliving Centre, University of New South Wales, Sydney).

An example of mixed cropping in a traditional swidden agriculture system. Upland rice, watermelons and jackbeans grow together in this family's seasonal garden plot in Rattanakirri, Cambodia.

Prior to planting the vineyard, the soil was conditioned by ploughing along the contour using a chisel plough to improve water absorption, aeration and root growth (courtesy of Nada and Tony Smark, Woodbrook Farm, Harcourt, Victoria).

An example of stacking plants to intensively use small areas, this Zone II orchard has groundcovers of sweet potato and herbs, an understorey that includes rosemary and small citrus trees, and a canopy of bananas and larger fruit-trees. Deciduous and evergreen trees are inter-planted, allowing sunlight through in winter for the chickens that live underneath (courtesy of the Permaculture Community Garden at the Ecoliving Centre, University of New South Wales, Sydney).

hazels. The prefix 'mel', from the Greek word for honey, on plant names is a useful indicator of bee-attracting flowers.

Pest management

Healthy soil and diverse habitats greatly reduce pest infestations. It has been found that in an apple monoculture up to 100 per cent of the apples can be infested with codling moth; in a polyculture only 4 per cent of the trees will be affected by codling moth. Pest management is achieved by:

- planting a diverse range of species so that no one pest can attack all the trees
- planting varieties with different harvest times; some varieties will miss the peak pest period
- allowing poultry and pigs to forage through the orchard; they will eat insect pests and diseased or infested fruit
- providing habitat for insectivorous birds and animals, including lizards, frogs and spiders
- maintaining a constant but not excessive supply of nutrients—over-fertilised, very lush growth attracts insect pests because it is highly palatable to them.

When you grow delicious fruit, many animals also want it. These are called raider animals and they will leave nothing for you. In Chapter 22 there are ideas for their management. In a small yard the size of a tennis court I would plant only dwarf trees and cover them with mosquito nets or enclose them completely with chicken wire. Trellis crops could grow on the fences, water could be supplied by a pond in the middle, herbs and bulbs could be planted as groundcover, and poultry would help to maintain the whole area.

Pruning

This is a contentious issue. Some people believe in it, others don't. Received wisdom says that you will always get the same total quantity of fruit from a tree: you can harvest many small fruit or fewer large ones. Good pruning is a very old skill. You must know what wood the fruit grows on—is it this year's growth or last year's? Old orchardists can shape trees so they don't grow long water shoots. I tend to prune once at the end of summer to keep the trees small enough to cover with nets after fruit set the following year, and they fruit heavily.

Plan seasonal activities one year in advance

By working with the seasons, you will find yourself using nature's patterns to enhance the productivity and health of your orchard.

Winter

- Fence, rip and implement your Zone II water-harvesting plan.
- Sow floral pasture seed.
- Order fruit to be planted in late spring.
- Plant new deciduous trees and pome, and take hardwood cuttings from fig, mulberry, grapes.
- Reduce chicken stocking rates if no groundcover.

Spring

- Plant nurse and windbreak species.
- Dig swales.
- Work on planting design for fruit trees.
- Sow green manure crop.
- Put chickens in to forage.
- Order citrus for planting in early summer and take soft-tip cuttings.

Summer

- Slash grass weeds at flowering.
- Sow autumn inter-crop species.
- Plant evergreen trees.
- Mulch all trees heavily.
- Put another beehive in or divide hive.
- Poultry can forage at quite heavy stocking rates to fertilise, control pests and eat any fallen fruit.

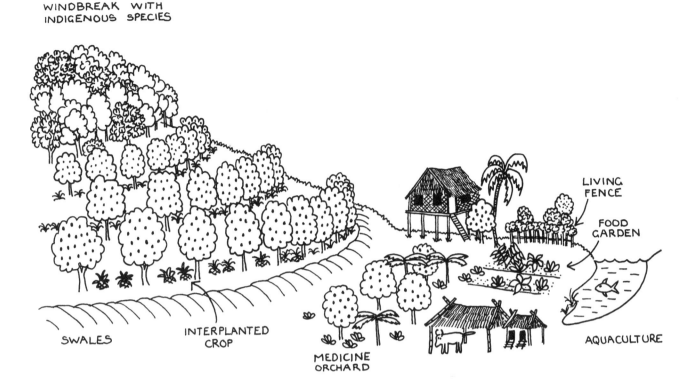

WINDBREAK WITH
INDIGENOUS SPECIES

SWALES

INTERPLANTED
CROP

MEDICINE
ORCHARD

LIVING
FENCE

FOOD
GARDEN

AQUACULTURE

Figure 13.8 Mountainous orchard in Vietnam. An example of an orchard in a hot wet climate.

Autumn

- Plant trees in winter rainfall areas.
- Plant trees in heavy frost areas while soils are still warm, and use tree protectors for the first year.
- Harvest fruit.
- Mulch heavily for winter.
- Remove or graze any seeding unwanted plants.

Retrofitting old orchards

Retrofitting is also called rolling permaculture. Existing orchards are usually monocultures or mixed orchards planted with only two or three species. In many cases they will be neglected, heavily infested with weeds, and will be unproductive, with many trees carrying dead wood. To renovate neglected orchards you can carry out the following steps.

Control weeds

Slash weeds and rip the ground with an agroplough, then put animals in to 'tractor' it (clean it up). Use different animals for different weed infestations: goats will clean up blackberry and wild roses; geese will eat grasses; chickens control oxalis, nut grass, kikuyu and couch. For ease of management, put the animals into one section at a time, confined by temporary fencing. On sloping land the fenced areas should be placed along the contours to reduce erosion.

Sow groundcover plants

After each section has been 'cleaned' and fertilised by the animals, sow an abundance of groundcovers, such as clovers, lupins, comfrey, buckwheat, turnips, radish, daikon, pumpkins, swedes, carrots and potatoes.

Establish windbreaks

Make sure you close the edge of the orchard to the prevailing wind. Choose productive windbreak species to provide bee fodder, fuel, bird habitat, mulch and additional animal fodder.

Pruning

An old orchard will not need pruning unless the trees are diseased, too large for easy harvesting, or prevent access. Remove all the dead wood and wait one harvest season, then mark the very best trees. Gradually remove the lowest yielding trees and replace them with new cultivars. Never plant more than 10 per cent of any one variety or species.

Orchards for all cultures

Permaculture orchards vary throughout the world. However, they share one principle in common: indigenous (native) fruit species are always included, either planted in windbreaks or used as regular orchard trees.

- In hot wet climates orchards look like tropical rainforests with the vertical stacking of many species.
- In cool temperate regions the orchard will resemble a deciduous forest with its floral pasture and broad spreading trees.
- In hot, dry desert areas the orchard will be relatively small and is designed to make the most of limited water. Tree species are typically deep-rooted to assist them withstand long dry periods, and are widely spaced to reduce competition for water.
- Often where land area is small, Zones I and II are integrated. See the diagram of the Vietnamese garden in Figure 13.8, which is how Zones I and II look in hilly climates of monsoon land.

Try these:

1. Sketch on your plan your ideal Zone II food forest site. Show the sunny aspects, how you will water the trees and where you will put the windbreaks. Now add all the species you have chosen. Remember to check the plants' traits so that you can place trees for optimum fruit set.

2. Write a list of the fruit you would like to grow. Eliminate the impossible; for example, pineapples in Tasmania, or blackcurrants in Singapore. Now make up a harvest calendar and fill in the cropping times so that you will have fruit throughout the year.

3. Work out a detailed seasonal timetable showing the stages of development of your orchard. Get your priorities right for the work and ordering or you may be forced to wait a year.

CHAPTER 14

The birds and the bees in the food forest

You have now begun to plant or renovate your orchard. This is the first stage. The second is to introduce animals to help you maintain it. Large and destructive animals are not appropriate in your food forest. Cows, horses, camels and elephants will compact the soil, break branches and raid the fruit. Small animals, such as poultry, bees and sometimes pigs, are much more suitable.

The benefits of having animals in your food forest include:

- They supply some of the orchard trees' nutrients.
- About 80 per cent of the animals' needs for food and medicine are met.
- They clean up weeds, insect pests, diseased and infested fruit.
- They 'tractor', or maintain, your orchard while supplying plant nutrients.

 ## Our ethical task is to:

- design and plant orchards which are integrated, diverse and permanent ecosystems of productive species providing habitat for plants and animals
- design water, fertiliser and soil improvement in orchards for long-term stability and low maintenance.

 ## Our design aims are to:

- know the needs of animals and how to meet them
- design healthy and humane housing

- use animals to support other enterprises
- introduce other small animals for their uses, if desired
- set up a hive of bees and provide for them.

 ## If we don't have design aims:

- our animals may not thrive and be healthy
- our animals won't carry out the functions we want them to
- orchard trees can be damaged.

Introducing animals to your food forest

How do you know whether an animal will integrate well? The analysis design method in Chapter 9 is most effective in achieving a smooth introduction of animals into your food forest. However, also ask yourself the following questions:

- Is the animal suited to the climate? Is there a locally adapted variety? What are your needs and tastes?

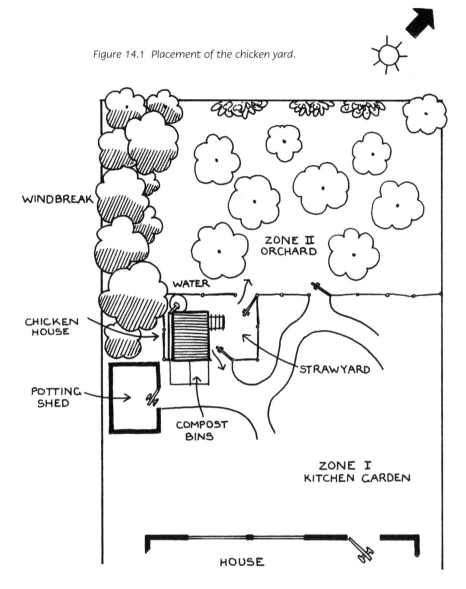

Figure 14.1 Placement of the chicken yard.

- What impact will the animal have on the environment? How will it integrate with other farm functions? What other uses does it have?
- How much space will the animal need? What are its husbandry needs? Who will take responsibility for it? What diseases is it susceptible to?
- What are the animal's breeding habits? If you don't want it to reproduce, what will you do?
- How does it interact with other animals?

In brief—what does the animal eat, what does it supply, and what does it do?

Chickens

Orchards and poultry systems are particularly synergistic because poultry, unless heavily stocked, do little damage. They are the best mobile species for your food forest.

Siting the chicken house

A chicken or duck house can become a polluter of the environment, so place it where the nutrient will run downhill into another garden or zone where you want the richer soils—a long way from the natural vegetation where you want the soils to remain natural. Chickens are mainly run in the orchard so the straw yard, or scratch yard, needs a gate leading into the orchard.

If you have a major weed problem in your Zone I garden it is also beneficial to occasionally let chickens into this area. Fence off a small area and let the chickens in there until the ground is scratched bare (tractored). The cleared area can then be rapidly replanted.

Figure 14.1 shows how access for chickens is planned for Zones I and II. The compost bins, propagation area (potting shed, cold frames, etc.) and straw yard should be placed in close proximity to the chicken house. The spent straw can then be easily collected and used for composting and mulching.

Housing needs

Chickens need warmth, safety, companionship and good health. Well-designed housing meets all of these needs. All construction timbers should be painted with sump oil and pyrethrum to prevent wood rot. Derris dust, lime and sawdust are needed on the floor of the nesting shed to repel lice, fleas and other chicken parasites. Figure 14.3 shows an example of chicken housing which meets most of the chickens' requirements:

- A water tank collects rainwater from the chicken house roof.

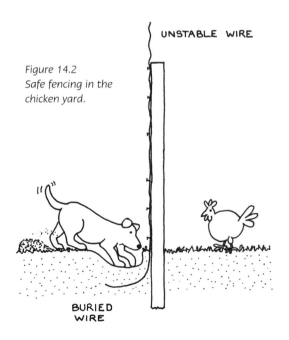

UNSTABLE WIRE

*Figure 14.2
Safe fencing in the
chicken yard.*

BURIED
WIRE

- Roosts are placed at the same height to stop competition, especially between roosters.
- Nesting boxes open from behind for egg collection.
- The insulated roof regulates the temperature inside the chicken house.
- The compacted manure floor of the nesting house builds up the chickens' natural antibodies.
- The straw yard is enclosed for protection. The straw yard also supplies fertiliser for plants.
- The fence is dug into the ground to prevent dogs, foxes and lizards digging under the wire (see Figure 14.2).
- The top wire is unstable and floppy to dissuade animals from getting in or out.
- The food tree in the straw yard deters hawks and eagles from raiding.
- Food crops and vines around the fences encourage foraging.

Small flocks are best left to range freely through the orchard. Figure 14.4 shows two designs suited to commercial chicken raising.

Breeds

White birds do better in hot areas while dark-coloured birds are suited to cooler areas. Rare or endangered breeds are particularly worthwhile because they tend to be more docile and more interesting animals than the over-bred commercial types. They also tend to live longer, are more flexible in their nutritional needs, and have a greater degree of acquired resistance to diseases.

Meeting chickens' needs

The food forest can meet 80 per cent of chickens' needs. However, it won't be able to do this unless you take the time to list the chickens' needs and yields, and decide how you will meet their requirements.

Medicines

Chickens will stay healthy if they have access to some of the following plants: oxalis, clovers,

TABLE 14.1: MEETING CHICKENS' NEEDS

Needs	Yields
Food—grits, grains, greens, grubs	Eggs
Water	Young
Company	Feathers
Disease control	Meat
Housing—safety, warmth	Pest management
Health—dryness, space, medicine, dust baths, friends and care.	Pleasure
	Company
	Warning system for predators
	Orchard maintenance

wormwood, mugwort and dandelions. Mugwort and wormwood can be grown in clumps and small hedges, and are said to repel lice and ticks. Onion weed, nut grass, couch and kikuyu are also good for general health.

Stocking rates

For all animals stocking rates are critical to maintaining both a healthy ecosystem and healthy animals. In permaculture we have two levels of stocking rates. Both systems are rotational and depend on your observations and design for the land and the animals.

● *Maintenance:* In this system, sufficient animals are stocked to browse and maintain the goundcover so that it is sustained without erosion, bare patches or damage to the land. The number of animals in an area has to be adjusted regularly to meet the seasonal conditons and feed supply. For example, during winter, or a dry season, the stocking rate must be reduced by removing animals or giving them a larger area to work.

● *Tractoring:* This actually creates bare ground temporarily in order to remove weeds and replant, or to remove the remains of a crop before introducing a second one. In this case you put animals in for a short time at high stocking rates until the ground is bare.

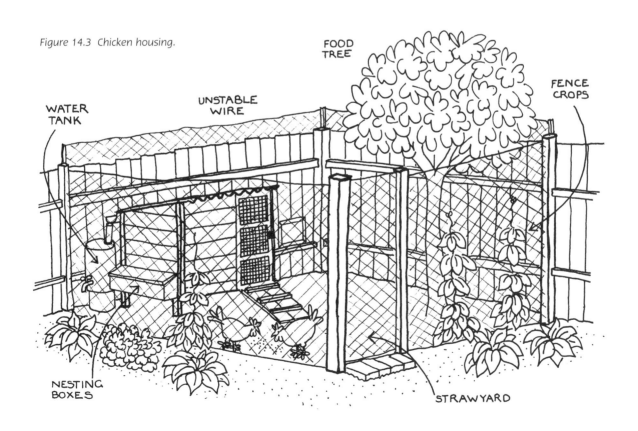

Figure 14.3 Chicken housing.

143

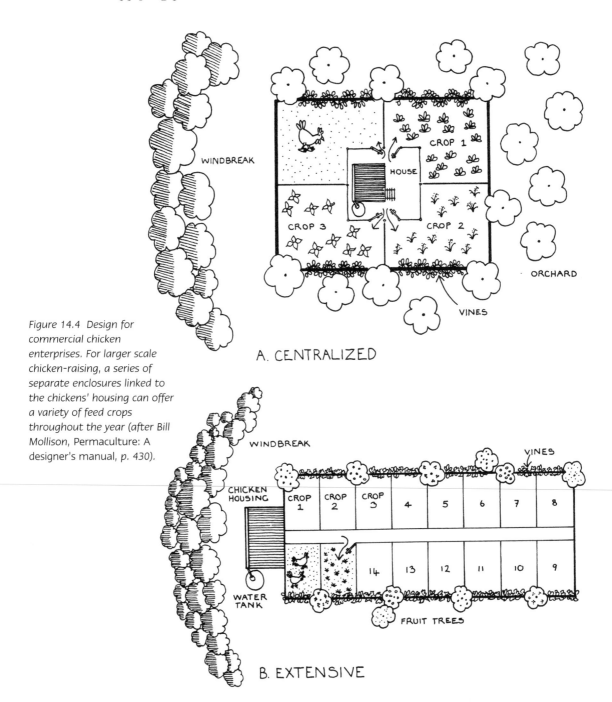

Figure 14.4 Design for commercial chicken enterprises. For larger scale chicken-raising, a series of separate enclosures linked to the chickens' housing can offer a variety of feed crops throughout the year (after Bill Mollison, Permaculture: A designer's manual, p. 430).

A. CENTRALIZED

B. EXTENSIVE

Depending on the season and breed, chickens raised in commercial numbers (50–100 hens) will maintain a half-hectare orchard very well. They can graze in fenced-off sections of 1/20 hectare, which are replanted each time they have tractored it.

Companionship

Chickens are sociable animals and need company. Their social orders are quite strict. Peace will reign if there are about 12–15 hens to one rooster; more

than 20 hens to one rooster and behaviour breaks down. However, at about 35 hens and two roosters, two flocks form. If there are two roosters they will both need to roost at the same height. If there are more than 35 hens and roosters to one pen group behaviour becomes aggressive and there will be fights in the yard.

Chickens prefer to nest at home and it is easier to find the eggs if they are let out at about midday after they have laid their eggs.

TABLE 14.2: FEEDING CHICKENS: THE FOUR GRS

GRs	Supplied how
Grains	Grow appropriate grains, either as part of the food forest groundcover or in a separate plot. Depending on your climate, you can grow grains from the following list: wheat, corn, amaranth, rice, millet, sorghum, barley, rye and oats. A mixture of grains is preferable. These grains are supplemented by seed from in-crop tree- and windbreak species, such as tagasaste, honey locust, carob, leucaena and acacias, and some groundcover seeds.
Greens	Plant them as part of your orchard's floral pasture and include the following: comfrey, clovers, chicory, oxalis, parsley, dandelions, clovers and herbs. The groundcovers are supplemented by fruit from vines on fences, such as grapes, chokos, passionfruit, kiwifruit, and windbreak species such as mulberries, hawthorn, elderberries, sunflowers, figs, guavas, loquats, tamarillo, pigeon pea and bananas.
Grit	Sand or crushed roasted eggshell assists the fowls' digestion. Usually, open-range chickens will find their own grit.
Grubs and insects	These supply the chickens' need for protein and will be plentiful in diverse and highly productive orchards. Bantams, being more insectivorous, are good at cleaning up fruit-fly larvae. All chickens enjoy a meal of termites. In fact if you make a run around a building, they will keep it termite free.

Other poultry

Ducks and chickens should be housed separately because ducks like wet, sloppy conditions and chickens prefer drier surroundings. Ducks are hardier than chickens. They can withstand colder, damper conditions and are less susceptible to disease. Ducks also eat more and are more efficient scavengers; however, they don't scratch the ground so the cultivating work done by chickens is not available. For maintenance of an orchard, temporarily fence off sections until the ducks have accomplished what you wish them to do. Ducks only require light, low fences.

They are generally less destructive in food gardens than chickens (although young seedlings must be covered) and are excellent for keeping down pests, especially slugs and snails. One Khaki Campbell will carry out good maintenance of the vegetable garden. Some breeds, such as Khaki

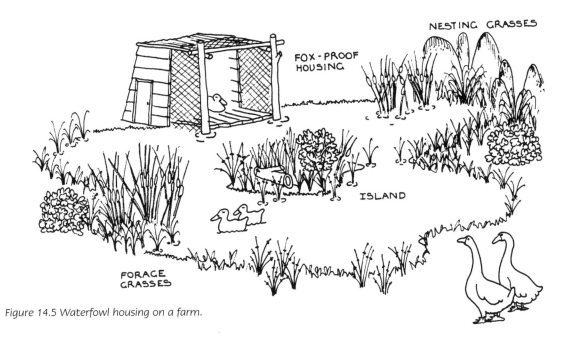

Figure 14.5 Waterfowl housing on a farm.

Campbell and Welsh Harlequin, will outlay chickens. Three duck eggs are equal to four standard-sized hen's eggs. A duck's laying life is two to three times longer than a hen's.

Their best habitat is an aquaculture system (see Chapter 24). Figure 14.5 shows suitable housing for ducks on a farm. Suburban dwellers can also keep ducks by simply providing a small pond or bath.

If you want to keep ducks, make a list of their needs, as we did for chickens, and then design a satisfactory system for them. There are no fixed rules so you can be imaginative.

Geese are the ideal animals for larger orchards since they are the 'grass weeders'. They will even weed between broad-leaved crops. Six to 12 geese per half hectare will maintain an orchard very well.

Pigeons can be kept in a dovecote and used for eggs and meat. They are visually attractive and have been used in traditional gardens for centuries. Their food requirements are grit, and larger seeds like maize, hemp beans and peas. Pigeon manure is excellent as a high-quality liquid manure, and is also a powerful insecticide—cover the floor of animal pens with pigeon manure to a depth

of 4 millimetres. Rats enjoy a good pigeon meal, so night housing must be rat-proof.

Bees in the food forest

Importing cane or beet sugar is not environmentally defensible. When grown under large monocultures, both crops are land degrading and polluting. Logically, honey is the best form of sugar. With a year's supply of honey, sugar can be taken off your shopping list.

It is not necessary to move the hives regularly to sources of pollen and nectar. Beehives with plenty of on-site forage can be permanently integrated into the site design. This is done by carrying out the same process of analysing needs and yields as we did for chickens.

Design to meet the bees' food needs

You need to know what plants are flowering throughout the year. There are plenty of bee-fodder calendars put out by apiarists' associations, which supply general information on flowering times. You can then adapt these to suit the plants and

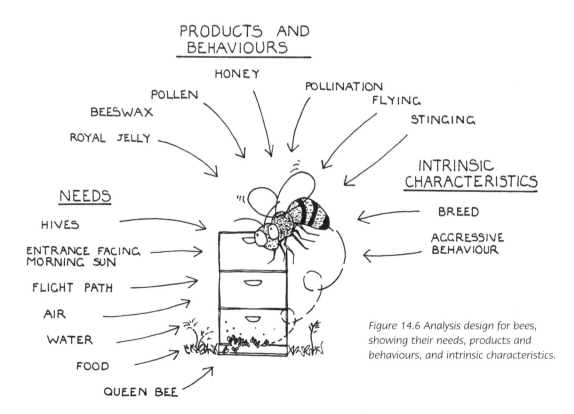

Figure 14.6 Analysis design for bees, showing their needs, products and behaviours, and intrinsic characteristics.

TABLE 14.3: MEETING BEES' NEEDS

Needs	Yields
Shelter	Honey—complex food
Food	Sugar with many complex minerals
Water—several times a day	Wax—ductile and high quality
Warmth—maximum sun	Pollen—high-protein food additive
Calm people	Propolis—natural silicon glue
Protection from wind	Royal jelly
	Pollination of countless plants
	Warning system for predators
	Broods

flowering times in your area and draw up your own forage calendar (see Figure 14.7).

In general, it helps to know that most berries and deciduous fruits are bee-pollinated, and that bees are attracted to blue flowers such as borage and lavender, and bulbs, including daffodils, iris, jonquils and freesias.

Bees will forage to a radius of 5 kilometres. Plant groups of forage plants so that the scout bees can easily find them and the workers have sufficient food for a week or more.

If possible, place the hives 50 metres from the most intensive forage areas. This gives nectar time to dry off on the way to the hive; otherwise it may form into alcohol.

Housing

There are many hive designs available, ranging from the old-fashioned curved hives to modern square boxes. You can make your own hives from patterns in do-it-yourself books.

A beehive contains a super and a brood box. The super contains the frames that bees make their honey in. The brood box is where the queen bee and her attendants work. Separate the brood box from the super or the queen bee will get into the super and feast on the honey.

The hive is placed about 1 metre off the ground to prevent predatory animals such as lizards, mice and cane toads getting into the hive. Position the hive to face east and make sure it's in an area where people won't need to walk past the bees' flight path. The roof of a building can be a good place for a hive because then the bees' flight path is above people's heads.

Shelter and warmth

Bees work less efficiently on cloudy and windy days and get cross, so place beehives in a sheltered position. In windy areas you will need to grow a windbreak. Spend time on your windbreak design to ensure that it also functions as a suntrap (see Figure 14.8). The windbreak species should also provide a considerable amount of bee food.

Bees like to wake up to warmth and sunlight, so ensure the hive opening faces east. In very cold climates, the hive should be insulated from the top, or one super removed from the top. I put bales of straw around the northern, southern and western sides of my beehive as insulation. This also acts as a suntrap.

Maximum solar gain is provided by placing the hive high up on a slope, away from trees and buildings. This is why you often see groups of hives placed on sunny hillsides some distance away from the trees that the bees are working.

Water

Bees must have a constant supply of clean, good-quality water. If they don't have reasonably accessible water they may have to search for kilometres and will make less honey. They can also die in droughts. A pond or in-ground bowl is suitable; however, one with a shelving edge is critical to prevent them from drowning.

Bee behaviour and human behaviour

Bees orientate themselves very precisely to the sun and dislike being moved. If the hives are shifted then the move must be 10 kilometres or more from their old home. For shorter distances move them no more than 3 centimetres per night. If the day is cloudy or windy, or if the beekeeper is feeling aggressive or upset, bees will pick up the agitation

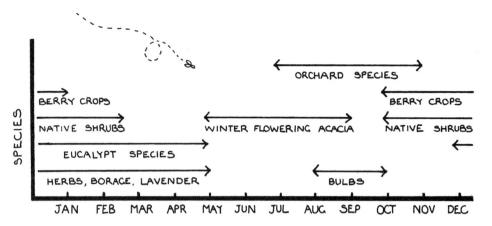

Figure 14.7 Example of an Australian bee forage calendar. A similar calendar can be made up for your area to show the local flowering time of bee forage plants.

and become upset and aggressive too. So have a calm presence around bees and always talk to your bees when you are in the orchard.

Bee stocks

There are several strains of honey bees, ranging from extremely mild and stingless bees to very angry, aggressive bees which will attack you even if you venture into the garden. If you have aggressive bees then this genetic trait can be changed by obtaining a queen bee from a more tranquil strain. Find out what is available in your area.

Honey

If you like special-flavoured honeys introduce clean frames when a specific plant begins flowering and then rob the hives again when the blossom is

finished. Some of the more desirable honeys are citrus, linden, eucalyptus, deciduous orchard, leatherwood, box, clover, blue gum and herb.

Try these:

1. On your site plan select a suitable site and design a chicken house. Make sure your design meets all the criteria for the chickens' safe and healthy housing.

2. List the additional things you want poultry to do for you other than producing meat and eggs. Are there jobs you really don't like doing in your garden and orchard? How can you use the animals' innate characteristics to work for you?

3. Where would you place a beehive and how would you supply food for the bees in winter and summer? On your plan draw the best place for the beehive.

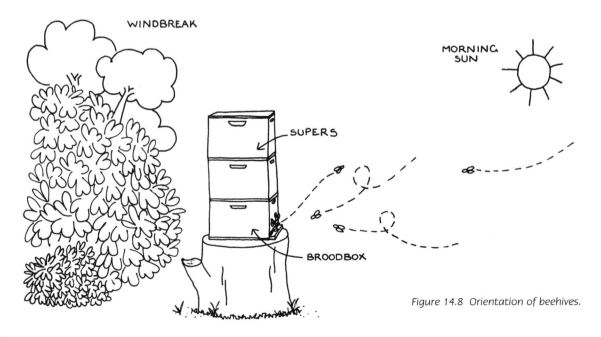

Figure 14.8 Orientation of beehives.

CHAPTER 15

If you are farming: Zone III

This chapter uses ethics and principles to help you to design a stable, productive farm that will endure extremes of drought and flood (the result of global warming). It introduces the principles of sustainable farming and offers some strategies which you, with imagination and experience, can develop into techniques.

Zone III is about design and strategies for traditional, commercial farming systems such as broad-scale cropping and foraging of any size in any climate. It is not about techniques. Lack of design skills has contributed enormously to rural degradation, which can be reversed by good design. The principles explored here will be equally relevant whether you already have a farm or plan to buy one.

Zones I and II are still integral to your farm plan because not only are they the most intensive areas of production, they add to your food and water security, carry out valuable ecological functions and help you survive if your primary enterprise fails. It is ironic that in Australia, in 2003, farmers caught by the failure of world prices in wool and wheat accepted food parcels from charities because they grew none of their own food. Usually there is a primary income enterprise; however, a secondary income must always be considered to reduce market and environmental risks. Remember, don't keep all your water in one tank.

On small land, or in cities or suburbs, you can combine Zones II and III. This can be done by incorporating a forage system for small animals (chickens, bees, ducks) in Zone II; growing a green manure crop for the fruit trees; or perhaps growing a small crop of barley or wheat. Look at Rob's

design in Figure 15.1 and see how he has combined Zones II and III. Land size is less important for income or self-reliance than intensity of production and good soils, water and protection.

Zone III is different in different climates and sites. Traditionally it is used for:
- staples such as rice, barley, oats, millet, corn, amaranth, sorghum, potatoes and rye
- larger scale organic fruit and poultry, or to increase enterprises from Zone II
- nuts
- raising market animals such as dairy cows, alpacas, llamas, sheep, goats, pigs, deer, fat lambs and, depending on the country, smaller indigenous animals such as wallabies.
- commercial crops such as rubber, oils, fibres, dyes and so on.

Zone III is not appropriate for very dry areas or areas with fragile soils that break down quickly under heavy cultivation or under adverse conditions. Some of the world's worst agricultural problems are caused by trying to intensively farm marginal land. It is better to leave these areas protected by natural ecosystems.

 Our ethical task is to:
- produce quality food without any environmental destruction

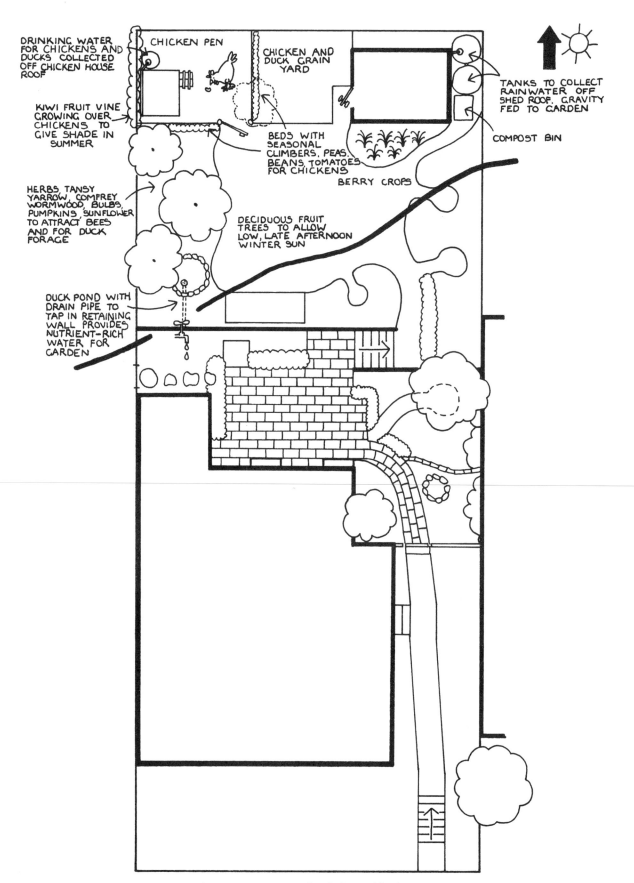

DRINKING WATER
FOR CHICKENS AND
DUCKS COLLECTED
OFF CHICKEN HOUSE
ROOF

CHICKEN PEN

CHICKEN AND
DUCK GRAIN
YARD

TANKS TO COLLECT
RAINWATER OFF
SHED ROOF, GRAVITY
FED TO GARDEN

KIWI FRUIT VINE
GROWING OVER
CHICKENS TO
GIVE SHADE IN
SUMMER

BEDS WITH
SEASONAL
CLIMBERS, PEAS,
BEANS, TOMATOES
FOR CHICKENS

COMPOST BIN

BERRY CROPS

HERBS, TANSY
YARROW, COMFREY
WORMWOOD, BULBS,
PUMPKINS, SUNFLOWER
TO ATTRACT BEES
AND FOR DUCK
FORAGE

DECIDUOUS FRUIT
TREES TO ALLOW
LOW, LATE AFTERNOON
WINTER SUN

DUCK POND WITH
DRAIN PIPE TO
TAP IN RETAINING
WALL PROVIDES
NUTRIENT-RICH
WATER FOR
GARDEN

Figure 15.1 Zones II and III design, Rob's place.

CHARACTERISTICS	RESULTS
BROAD-SCALE FORAGE SYSTEM.	• TREES, SHRUBS AND GRASSES PROVIDE FORAGE FOR LARGE GRAZING ANIMALS (COWS, SHEEP, GOATS, ETC.).
TREES GROWN IN ROW CROPS AND WINDBREAKS.	• MAJORITY OF SPECIES PROVIDE FORAGE FOR GRAZING ANIMALS. • SOME PLANTS YIELD FIREWOOD, MULCH, EDIBLE FRUITS AND NUTS ETC.
SUSTAINABLE CULTIVATION TECHNIQUES.	• GRASSES ARE SLASHED AND TREE BRANCHES ARE CUT TO PROVIDE MULCH WHICH PROTECTS AND BUILDS THE SOIL.
SPOT MULCHING.	• INDIVIDUAL YOUNG TREES ARE MULCHED TO ASSIST FAST ESTABLISHMENT AND GROWTH.
INTERCONNECTED DIVERSION DRAINS AND DAMS THROUGHOUT THE ZONE.	• HIGH DAMS AND LOW DAMS ARE CONNECTED BY DIVERSION DRAINS TO PROVIDE WATER TO ALL PARTS OF THE ZONE. • GRAVITY FEEDING ALONG DIVERSION DRAINS PROVIDES A LOW MAINTENANCE WATERING SYSTEM. • DAMS INCREASE HABITAT DIVERSITY (THROUGH MICROCLIMATE CREATION). • DAMS AND SWALES ENSURE THE LAND IS PROTECTED AGAINST DROUGHT.
DIVERSE HABITATS FOR WILDLIFE AND DOMESTIC ANIMALS.	• HABITAT DIVERSITY INCREASES SUSTAINABILITY AND STABILITY. • ANIMALS CONTROL PESTS.

Figure 15.2 Characteristics of Zone III.

- reverse land degradation and water depletion.

Our design aims are to:
- place Zone III appropriately on our land
- protect it from soil, water, wind erosion and pollution
- prepare the zone for commercial cropping.

If we don't have design aims:
- land and enterprise will be vulnerable to market and climate instability
- land will reject us—how many farmers have walked off their land because of lack of design skills?
- costs will be inflated.

Site selection

Zone III is usually larger than Zones I and II and connected to them in some way. With flat land it would be the third concentric circle from your house. However, with different site topography the model is altered to match the land.

Know the needs, yields and resilience of the enterprise, then make sure you can meet its needs

TABLE 15.1: CRITICAL CONSIDERATIONS FOR ZONE III

Critical considerations	Design strategies and principles
Water	You will need a water-harvesting scheme that allows water to service the whole area, preferably through gravity flow. Water is needed for stock and crops. On large sites capture it in high dams and then gravity-feed along interconnecting swales to lower dams. The swales placed along the contours help replenish soil moisture and absorb excess in times of heavy rainfall. Plant the tops of hills, creek lines and riverbanks.
Access	Easy vehicular access is required for harvesting, and for managing animals. Select the higher contours to build the access roads so you can view your property, and always use the same roads or tracks to prevent vehicles compacting the soil.
Nutrients	These must be supplied in the right quantities, form and season, and must not pollute. You choose fertilisers from alley cropping, animal manures, nitrogen-fixing trees, clover and green manure crops.
Slope	Land with a slope greater than 15 degrees should be terraced to prevent erosion, or grow only permanent tree crops and have light stocking rates.
Fencing	Design your fences thoughtfully to manage movement of animals and feed. Use electric fences with solar panels for energy. Try to place long fences along contours and not down slopes where animals will wear erosive pads.
Aspect	Animals need sun and wind protection in summer and winter. Some crops need easterly or westerly aspects. Study your site to place crops where they will thrive.
Protection	Natural features such as hills and forests will provide protection from the prevailing winds. However, you need to carry out a sector analysis and then plant mixed-species windbreaks for multi-harvest products and functions. Preserve and extend any natural forest and vegetation.
Soils	Are not ploughed or inverted. Fertility is constantly built up.

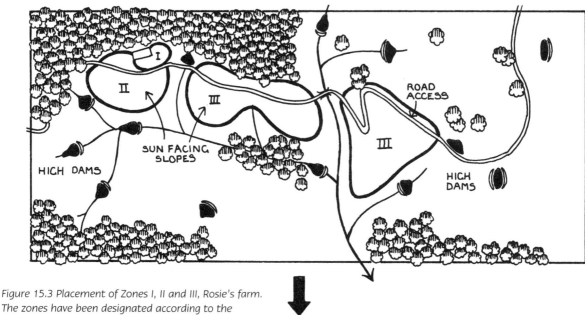

Figure 15.3 Placement of Zones I, II and III, Rosie's farm. The zones have been designated according to the topography and to the characteristics of each zone (i.e. their yields, functions and maintenance requirements). Zone III requires less maintenance than Zones I and II, so it has been placed farther away from the house.

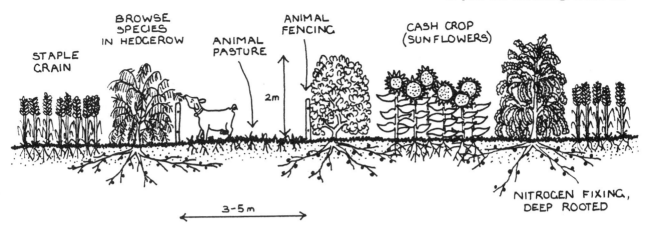

Figure 15.4 Profile of hedgerow intercropping (after B. Mollison, Introduction to Permaculture, p. 30).

in these critcal factors and never degrade or pollute the environment.

Look at the plan of Rosie's farm (Figure 15.3) to see how she has chosen her Zone III with these factors in mind.

Alley cropping

Also known as hedgerow intercropping, alley cropping is the farming technology proving to be most effective and productive for Zone III. It is characterised by a permanent structure of trees inter-planted with arable crops. This farming technique is called hedgerow intercropping or alley cropping because the trees are grown in wide rows and the crops, or animals, are placed in the interspaces or 'alleys'.

Advantages of alley cropping

- Increased crop yields due to added nutrients and organic matter.
- A reduction in, or elimination of, chemical fertilisers.
- Improved nutrient recycling because the tree

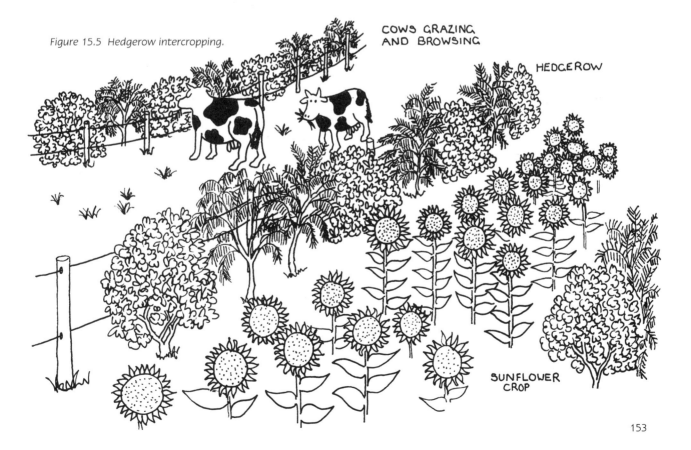

Figure 15.5 Hedgerow intercropping.

TABLE 15.2: SPECIES FOR TREE ROWS

Criteria for tree selection

Ease of establishment from seed or cutting
Rapid growth rate
Good coppicing potential
Deep-rooted—crop roots usually shallower
Multiple uses—firewood, forage
High leaf–stem ratio
Small leaves or leaflets
Withstands pests and diseases

Species used successfully

Tropical and subtropical
Legumes: *Leucaena, Glyricidia, Cassia, Erythrina,
  Tephrosia, Sesbania* species.
Non-legumes: *Acioa, Alchornea, Gmelina* species.
Temperate
Tagasaste, Chamaecytisus, mulberry, poplar, etc.

rows exploit moisture and nutrients deep in the soil and not available to most crops.

- A physical improvement in the soil because the trees modify soil fluctuations and reduce soil moisture loss. There is also better water infiltration and reduced run-off.
- Less soil erosion on sloping land because the trees act as physical barriers to soil and water movement.
- Extra yields such as forage, firewood and timber.
- Healthier animals because they have a wider dietary choice—tree fodder as well as grasses.
- Improved weed control—first by shading, and later from mulch.

Disadvantages of alley cropping

- Possible competition between trees and crops for light, water and nutrients.
- A reduction in crop area due to tree rows.
- Additional labour is required to establish the system.
- Some limitation in flexibility of land use.

Overall, the merits of alley cropping appear to heavily outweigh the disadvantages.

Preparation

Allow the system to establish before putting in crops and animals because time spent on preparation will pay off. Regard this stage of development as an investment.

- Before planting develop water systems and plan the placement of your dams and swales in relation to the topography of the site and planned enterprises.
- Establish windbreaks early so your crops and animals will have protection from weather extremes. A three-row windbreak is the minimum; some areas require a five-row windbreak. Select species carefully so the windbreak provides additional yields such as nuts, timber, mulch and fodder, and has the multiple functions of a wildlife corridor, firebreak and soil protection.

Tree rows, the hedges, are usually established along the contours at 3–5-metre intervals. The width of the alleys depends on how much space you want to devote to crop growing, how much space is required for growing trees, and the soil and climate. Spacing commonly used between rows is 2–5 metres.

The trees are regularly pruned to provide mulch, which is used to improve soil organic matter and provide nutrients, especially nitrogen, to the crops. For this reason, it is beneficial to include nitrogen-fixing plants in the hedgerows: the leaves of these plants are high in nitrogen, and when dug into the soil, will raise the nitrogen level. In addition, their roots 'fix' nitrogen (see Chapter 8 on plants).

Alley crops

Most cereal production adapts to alley cropping and the system is flexible enough to allow different types of crops to be grown each year, without many problems. For example, barley could be grown one year; the following year rye could be planted in the same area. Crops that have been successfully grown include maize, sorghum, cassava, barley, upland rice, sunflowers and pineapples.

Row trees mulched for the alley crop

The timing and frequency of pruning depends primarily on the tree species (in relation to the

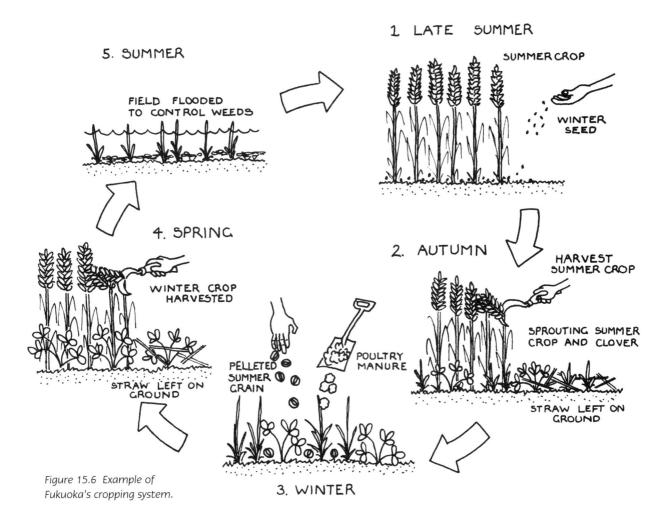

5. SUMMER

FIELD FLOODED
TO CONTROL WEEDS

1 LATE SUMMER

SUMMER CROP

WINTER
SEED

4. SPRING

WINTER CROP
HARVESTED

STRAW LEFT ON
GROUND

2. AUTUMN

HARVEST
SUMMER CROP

SPROUTING SUMMER
CROP AND CLOVER

STRAW LEFT ON
GROUND

PELLETED
SUMMER
GRAIN

POULTRY
MANURE

3. WINTER

Figure 15.6 Example of
Fukuoka's cropping system.

trees' growth rate and height) and on the crop's lifecycle. A well-established tree has greater reserves of nutrients. The time of cutting is usually six to 12 months before the first crop is planted, and in general you can expect to cut the tree rows about three times during a growing season of six months.

A low cutting height is desirable, and 2 metres is the maximum height for the row trees. If taller than this, they will have a shading effect on the alley crop. The trees are pruned with a slasher; the prunings are then fed into a mulcher and chopped up into fine organic material. The mulch can be applied as a surface covering which helps to control weeds and assists water retention in the soil, or it can be incorporated into the soil to improve the efficiency of nutrient transfer to the crops.

Row trees browsed by animals

Animals can either browse directly on the trees or the harvested leaves can be taken to a separate area where the animals are kept. If you allow the animals to browse on the trees, they will need to be controlled by electric fences (to prevent them overgrazing the trees). The stocking rates will be dictated by their rate of eating. Animals should never be allowed to eat all the leaves as the plants will not recover from this. In a small area it may be better to cut and carry the mulch to the animals.

Fukuoka's method of natural farming

Cereals are successfully grown in Zone III in cool temperate areas where the soils hold large quantities of organic matter. In the tropics they are

155

disastrous, with one exception: when rice is grown with ducks and fish.

In Japan, Masanobu Fukuoka, a plant pathologist, returned to his native village and noticed that rice growing wild in neglected fields and alongside roads had ears as full as those grown under cultivation and chemical loads. After close observation, Fukuoka then copied the conditions exactly and successfully grew healthy rice among the 'weeds' without disturbing the soil with machines or using chemical fertilisers. His radical approach was called Fukuoka's Method of Natural Farming.

Fukuoka's four main principles

1. *No cultivation:* Ploughing and other methods of cultivation which disturb the soil are not used. Instead, natural cultivation occurs through penetration of roots and the work of soil organisms and small animals.
2. *No chemical fertilisers or prepared composts:* Fukuoka found that soil left to itself naturally maintains fertility in accordance with the orderly patterns and cycles of plant and animal life.
3. *No weeding by machines or herbicides:* Weeds play their part in building soil fertility, therefore they are controlled but not eliminated. Straw mulch, groundcovers such as white clover (nitrogen-fixing), and temporary flooding provide effective weed control.
4. *No dependence on chemical pesticides:* Harmful insects and plant diseases are always present yet do not require chemical control if natural balances are maintained. The best control is to grow sturdy crops in healthy soil.

An example of a Fukuoka cropping system

The timing of seeding is such that there is no interval between succeeding crops. Essentially, this is stacking in time and space and it gives grains an advantage over the competing weeds. The timing requires specialised knowledge of local microclimates; for example, when to expect frosts, opening rains and seed germination.

In *late summer*, seeds of the winter crop (rye or barley) are sown while the summer crop (rice) is still ripening in the field. Clover is sown with the grain to provide nitrogen and to help keep weeds under control.

The summer crop is harvested in *autumn*. The straw and chaff from the crop are returned to the field to provide mulch. The winter grain begins to germinate through the mulch.

In *winter*, poultry manure is scattered over the field, or poultry are let into the field to add manure and clean up pests. The summer grain is pelleted (that is, the seed is rolled in clay—this protects the seed and prevents birds eating it), and is sown in the fertilised field.

In *spring*, the winter grain is harvested and the remaining straw and chaff is returned to the field. The summer crop is now starting to germinate and the clover is growing strongly.

In *summer*, the field is flooded to suppress the clover, to prevent summer weeds growing, and to help the rice grow.

Growing nuts in Zone III

Nuts are a high-priced and easily stored crop. They are best grown in a low-maintenance orchard or in an alley cropping system. Like other trees in this system, they can be regularly pruned to provide soil nutrients and mulch. Eventually the other trees in the rows will die out as the taller-growing nut tree canopies exclude light.

The most commonly grown nuts with good market prospects are:

- almonds
- hazelnuts
- pine nuts
- bunyas
- macadamias
- pistachio nuts
- chestnuts
- pecans
- walnuts.

Cashews are allelopathic—they secrete chemicals into the soil that inhibit the growth of other plants. They grow best in monsoon climates and relatively poor soils.

Hazelnuts are a particularly good windbreak species for cool climates. They have a fine twiggy habit which effectively filters dust and wind in summer; in winter, they lose their leaves and allow sunlight through the canopy. Several species of nut trees, including pine nut, walnut and chestnut, have extremely valuable timber. In the event of the nut crop failing, the timber provides a back-up income.

Animals in Zone III

Most people in the Western world think of animals as suppliers of food products or as pets. However, until the advent of the Industrial Revolution and the exploitation of fossil fuels, most people were dependent on the muscle power of animals. In many parts of the world, draught animals are still vitally important for food production and transportation. And, because of the high capital and operating costs of machines, working animals are likely to remain essential power bases of the developing world's small farmers.

In rich countries, animals are often kept in inhumane intensive systems where even their previously valued by-products, such as manure, are seen as problems. The animals' diet is often unnatural and lacking in variety, which can cause health problems. Organic farmers report much lower veterinary bills than conventional farmers. In Zone III, where grazing animals eat a wide range of food plants and stocking rates are kept fairly low, animals stay healthy and require veterinarians much less frequently.

Choosing animals

Be imaginative about the animals you choose. As well as obvious products such as food, meat, milk and eggs, animals can be chosen to provide weed control, fertiliser (manure), soil cultivation (digging and scratching), security (geese are excellent watchdogs), work (when trained to harness), and for by-products such as skins, fibres, feathers and transport. In addition, you may decide to keep animals which are rare or endangered so they can be reintroduced later into their natural ecosystems. You may also decide to breed heirloom or threatened varieties of animals so the gene bank can be preserved.

The natural environment will influence your choice of animals. Hard-hoofed animals will compact heavy clay soils; if you really need these animals, select a smaller breed and carry them at lighter stocking rates. If your land borders natural ecosystems it must be well fenced to prevent the animals escaping and becoming pests.

Animals are a natural part of the cropping

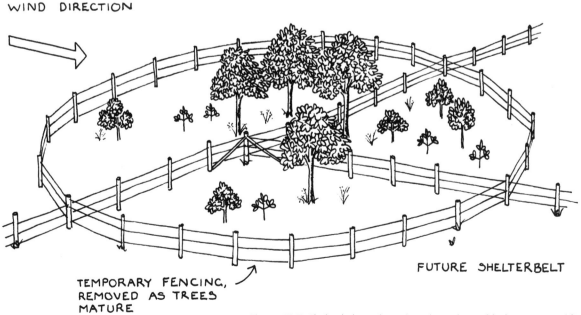

WIND DIRECTION

TEMPORARY FENCING, REMOVED AS TREES MATURE

FUTURE SHELTERBELT

Figure 15.7 Shelterbelt at the point where the paddocks meet provides animal protection and seed distribution for self-sown windbreak.

system and should not simply be regarded as a means to an end. They have needs that must be met; if these are satisfied then the animals will produce well and stay healthy.

Animal needs are:

- *Food:* Grow a range of animal fodders according to your site and climate, and encourage animals to perform weed control. Generally, most animals graze (they eat grasses) and browse (they eat trees and shrubs). Animals with a varied and mixed diet get sick less often and require fewer veterinary visits.
- *Water:* Water must be clean and freely available. It must also be close to where the animals feed and seek shade. Many small watering points are better than one large dam. In very hot weather, supplementary fodder such as windfall apples helps to relieve the need for water, as well as supplying vitamins and minerals.
- *Shade:* In very hot conditions animals need shade. No animal should ever have to walk more than 400 metres to find shade.
- *Protection:* Windbreaks act as shelterbelts in cold and windy conditions, as well as a source of supplementary feed (see Figure 15.7).
- *Housing:* Animal housing should be easily accessible from your home. Factors influencing siting are much the same as for human housing: animals require light, warmth and protection from winds. Doors and windows should face the morning sun, and windbreaks and shade over the structures will reduce animal stress. Many animals need a holding yard as well as stabling; in this case large shade trees are

necessary. The holding yard is also a good source of manure.

- *Treatment yards:* Yards for milking and shearing are placed close to the home. The yards should be positioned upwind from the house because dust and animal sheddings can be blown into the house and may cause asthma or other health problems. Windbreaks should be grown around the yards. Fodder trees can also be grown on the periphery for supplementary grazing— animals held for treatment are more placid if they can graze.
- *Parasite control:* To break the cycles of eating and excretion of worms and eggs, animal cropping areas need to be rested. This is known as rotational grazing. There are different times for different parasites and often quite good control can be obtained by moving the animal or the pen around the property, thus saving the need for chemical drenches.

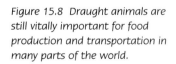

Try these:

1. If you own a farm, site Zone III on your site plan. If you do not have a farm but would like to try farming, draw an imaginary farm that you feel would suit you and put Zone III on it.

2. Write down the types of enterprises you would like. (Be realistic about your ability to manage them.) How would you fit your enterprise into a hedgerow intercropping system?

3. Now select the plant and animal species you would include in Zone III. Specify the amount of time you will need to establish the enterprise and the space you will require.

Figure 15.8 Draught animals are still vitally important for food production and transportation in many parts of the world.

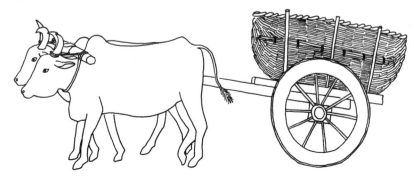

CHAPTER 16

Harvest forests: Zone IV

For hundreds of years people have planted trees to serve the needs of future generations for timber and tree products. In Europe and Asia forests were planted by rulers, by churches and by city authorities. By planting trees they practically and symbolically showed their faith and hope in the future. They planted trees for timber and for non-timber products (NTPs), such as fruit, seed, dyes, oils, barks, mulch and other products that can be harvested without the trees being cut down.

Throughout eastern and southern Asia people have always declared some trees sacred and burnt incense at their base in gratitude for all they do. In Tasmania giant ancient forests hundreds of years old are felled for cheap woodchips. Forests strongly exemplify our attitudes to nature and the future.

Trees are fundamental to the life processes that maintain clean air, water and healthy soil. The biologist E.O. Wilson in his book *Biodiversity* gives evidence that in regions where tree cover drops below 30 per cent of the original forest, other sustainable life processes begin to collapse. Rivers silt up, soils wash away and air quality declines. James Lovelock, the renowned scientist, claims that these breakdowns of natural systems in turn affect the other world bio-areas such as deserts and polar regions; for example, monsoon rains fail on another continent or cyclones occur more frequently. We can now see this happening. Jared Diamond in his book *Collapse* provides cogent evidence for the same destabilisation.

In this chapter small and large forests are examined as a farm enterprise for their products. To harvest the products, tree removal is minimised. Every suburb, city and farm needs a forest to supply its own needs and carry out ecological functions.

Enough trees need to be replanted to account for all the timber and NTPs used by every person—like putting money back in the bank.

Plant a forest if you have access to rural land or, if you live in the suburbs or a city, seek out where government or council land is available. It is important that everyone tries to plant back all the trees that they have used throughout their lives. Government forestry departments are not doing it. How many trees are needed for all the paper, all the firewood, all the houses, all the furniture and all the tree products that one person uses? You need two substantial trees per person per day for the rest of your life.

Perhaps you have noticed that in Zones II and III you are also increasing your permanent tree and shrub cover with orchard trees, nut trees, windbreak trees, tree rows and shelterbelt trees. You are stabilising your environment, increasing diversity and becoming more self-sufficient.

 ## Our ethical task is to:
- increase all forests everywhere
- ensure that when a tree is harvested it is replaced, and that its products last as long as the tree took to grow.

CHARACTERISTICS	RESULTS
LONG-TERM TREE PLANTINGS.	· TREE PLANTINGS PROVIDE HARVESTABLE PRODUCTS FOR PRESENT AND FUTURE GENERATIONS.
DIVERSE YIELDS AND FUNCTIONS.	· PLANTINGS GIVE MANY PRODUCTS AND YIELDS EG. DYES, PAINTS, OILS, BEVERAGES, NUTS, FLOWERS, SEEDS, BARK, ROOT PRODUCTS, MULCHES, FIBRES, HONEY, MEDICINES, POISONS, FUEL, FIREWOOD, BUILDING TIMBERS, FOR POLES, FURNITURE, PAPER, BASKETS AND BOATS.
MULTIPLE BENEFITS FOR THE WHOLE LAND.	· THE STRUCTURAL FOREST INCREASES SOIL MOISTURE; MAINTAINS THE WATER TABLE; CLEANSES WATER; REDUCES EROSION AND DESICCATION; PROVIDES A STORE OF GENETIC DIVERSITY; ACTS AS A WILDLIFE CORRIDOR; AND PROVIDES FORAGE FOR GRAZING ANIMALS · THESE BENEFITS EXTEND BEYOND ZONE IV TO THE REST OF THE PROPERTY AND ADJOINING LAND.
UNDERSTOREY GROUNDCOVER PLANTINGS.	· CLOVERS AND INDIGENOUS GRASSES PROVIDE ANIMAL FODDER, AND BUILD AND PROTECT THE SOIL.
MAINTAINED BY LARGE, HARDY, RANGE ANIMALS.	· GRAZING ANIMALS ARE KEPT AT LOW STOCKING RATES TO CONTROL WEEDS, PROVIDE FERTILISER, ETC.

Figure 16.1 Characteristics of Zone IV.

TABLE 16.1: PLACES FOR FORESTS

Where	Description
Urban forests	By co-operating to plant urban forests, people in cities and towns can begin to meet their own needs for trees and tree products in the future, plant trees to give back some of the tree products used in their lifetime and to take the stress off plantations and natural forests. Derelict land can be rented for the purpose; it will become a fun and joyful place to be as the trees grow.
Suburban forests	These can be planted in schools, parks, hospitals, near streams and along roads. Plant street by street, or ask your local government body for land to develop as signs of hope for the future.
Farm forests	Site your harvest forest on land not suitable for orchards or farming, or on steep or shady slopes. Look at Rosie's farm (see Figure 16.2) and you will see how harvest forests naturally encircle farming in Zone III and Zones I and II. In the event of hard times or crop failure, a harvest forest protects the water catchment and supports cropping areas.

Our design aims for trees are to:

- keep 30–35 per cent of all land permanently tree covered
- select species carefully for their yields and functions
- harvest on a long-term sustainable basis for timber or their products
- not harvest trees without replacing them
- plant enough trees to give back to future generations the timber and tree products we have used and not replaced.

If we don't have design aims for trees:

- desertification will continue unchecked
- all drinking waters will become polluted, silt up and, in time, dry up
- climates will be further destabilised
- microclimates will become harsher
- tree products will become increasingly rare or disappear.

Siting Zone IV

This zone is usually contiguous with Zone III and Zone V, and sometimes with Zones I and II. Figure 16.2 shows how Zone IV is sited on Rosie's farm.

You can also plant trees as a crop in Zone III, and in Australia farm forestry has the same taxation status as other rural enterprises. In this case you will have a food garden (Zone I) and orchard (Zone II), and Zones III and IV with trees as the commercial crop.

Site considerations

Before you rush out the door with a seedling in one hand and a spade in the other, go back to your site inventory and walk over your land again. Now consider the site factors for Zone III in Table 16.2.

Establishment

Like all the other zones, your enterprise will be more successful if water systems and swales are developed before planting. Establish your wind-breaks with their edge to the prevailing wind and, remember, do not break the edge (see Figure 16.3).

Farm forestry options

If you want to farm trees as your primary enterprise then you have two options:

- Farm forestry with domestic animals to maintain and fertilise the trees.
- Farm forestry with indigenous animals for maintenance and fertiliser.

TABLE 16.2: SITE CONSIDERATIONS

Factor	Consideration
Labour	Tree planting is labour intensive. Will your neighbours co-operate and swap labour? Do you need a nursery to grow seedlings? Or can you order them and how soon in advance of planting?
Access and machinery	Specialised machinery is expensive. Will your Landcare group invest in it? Is hiring better than owning? Are all the roads along the contours?
Slope	Excessive slope may need to be terraced or swales may need to be made.
Water	Are there high sources of water to gravity-feed, and is there enough water for establishment of crop
Soils	Some trees are fussy. Do you know their needs? What will you do about nutrients for the growing crop? Consider your options.
Protection	Trees need to be protected from frost, animals, people, drought, etc. until they can look after themselves.
Desired yields and functions	What do you want the trees to provide: quick-growing timber, cabinet timbers, NTPs of a huge range, stabilisation of soils, protection of water, rivers and slopes?

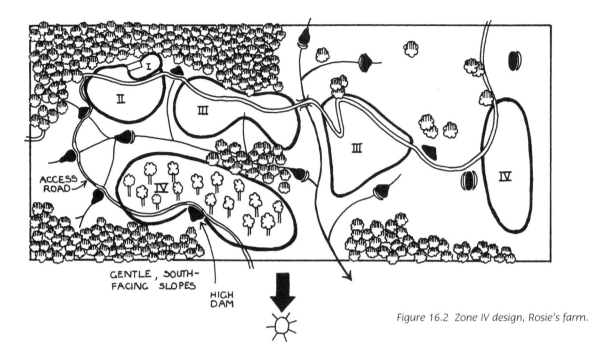

Figure 16.2 Zone IV design, Rosie's farm.

Farm forestry and domestic grazing animals

Areas from 2 to 16 hectares will support a self-maintaining forest. Smaller areas will need more work to keep them weed free and watered.

The first trees planted are nitrogen-fixing and pioneer species which act as 'nurses' for the later planting of climax seedling species. These first plantings include fodder trees for the grazing animals. The climax tree species are chosen for their yields and suitability for your environment. Many valuable timber trees will not grow in a monoculture because they require a companion shrub or tree species (often a nitrogen-fixing species). You can obtain lists of timber trees that require companions from your local Forestry Department or National Parks office.

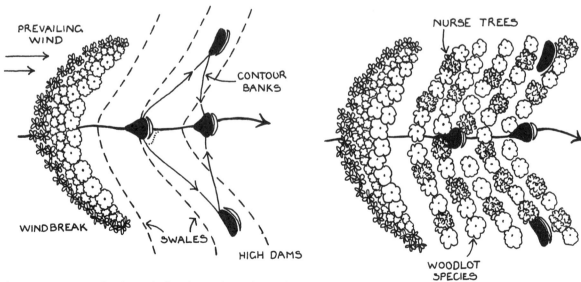

Figure 16.3 Zone IV planning. Windbreaks, swales and watering systems should be established before the structural forest is planted.

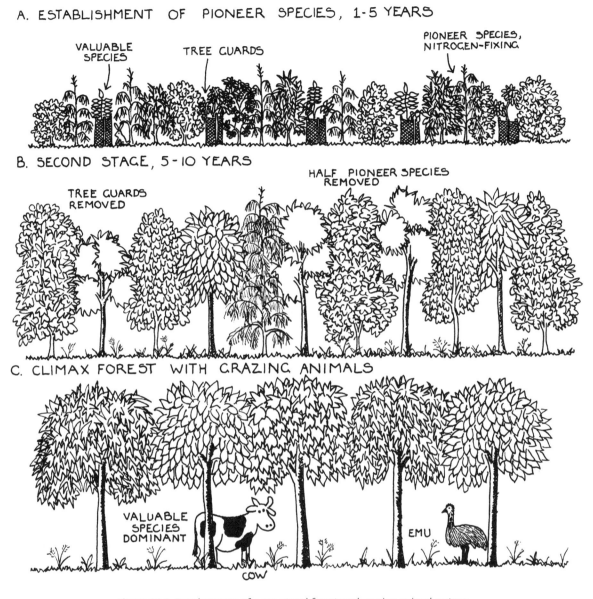

A. ESTABLISHMENT OF PIONEER SPECIES, 1-5 YEARS

VALUABLE SPECIES

TREE GUARDS

PIONEER SPECIES, NITROGEN-FIXING

B. SECOND STAGE, 5-10 YEARS

TREE GUARDS REMOVED

HALF PIONEER SPECIES REMOVED

C. CLIMAX FOREST WITH GRAZING ANIMALS

VALUABLE SPECIES DOMINANT

EMU

COW

Figure 16.4 Development of a structural forest and grazing animal system.
As the forest matures, pioneer species are gradually removed to allow productive species to develop.
Selected grazing animals can be introduced to maintain the forest floor.

The trees are planted closely together in lines and in widely spaced rows. This encourages the tree trunks to grow straight. You will need to take into account the canopy diameter of the mature trees. Most good catalogues and tree books give these figures.

Grazing animals are introduced into the forest as the trees mature—from three to six years depending on climate and species. Grazing is controlled through light stocking rates. By this stage the short-lived pioneer species will have either died naturally or will

have been harvested for their short-term yields, such as mulch, poles and firewood. Animals such as beef cattle, emus, large deer, llamas, ostrich or wool sheep graze the grass and clover understorey, which grows well under the filtered light of the trees. Figure 16.4 shows how your forest develops.

At maturity 500 to 1000 high-value trees per hectare form your structural forest and support grazing animals. Every year you harvest, replace and replant to keep this number of permanent trees forever.

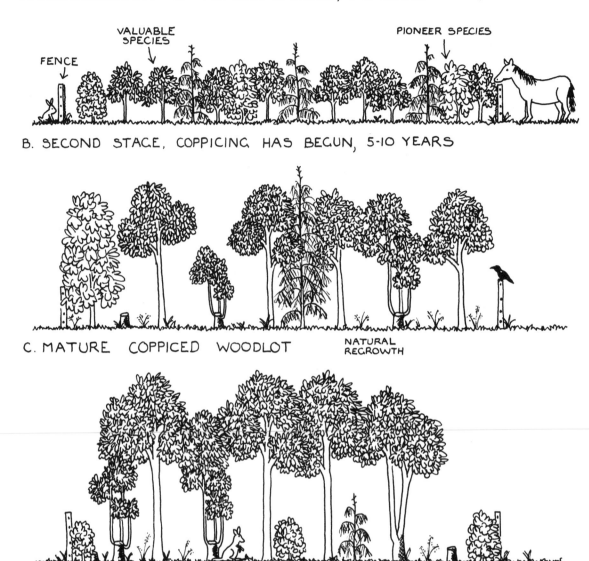

A. ESTABLISHMENT OF PIONEER SPECIES, 1-5 YEARS

FENCE VALUABLE SPECIES PIONEER SPECIES

B. SECOND STAGE, COPPICING HAS BEGUN, 5-10 YEARS

C. MATURE COPPICED WOODLOT NATURAL REGROWTH

Figure 16.5 Coppiced woodlot with animals.

Farm forestry and indigenous grazing animals

Again, start with windbreak and water systems. Plant so that every two years you will cut or coppice one row of trees. Coppicing involves cutting a tree at about chest height; after several years, it grows many new limbs (see Figure 16.5). Coppiced trees give you firewood, mulch, etc. on a renewable basis without having to fell the entire tree. By removing some species the final climax species will develop without stress or competition for light or nutrients.

Young trees must be protected from rabbits and grazing animals, but once the trees are tall enough, the local indigenous animal species such as deer, ostrich, emu, kangaroo, buffalo, elk, or llamas maintain the groundcover. After the trees are established these animals roam freely to maintain forest functions.

Table 16.4 is a climate guide to selecting long-term fine wood species. These are the main types of forests, and their harvesting and replanting needs to continue forever.

TABLE 16.3: HARVEST FOREST YIELDS, SPECIES AND MANAGEMENT

Yields	Criteria and species	Management
Firewood	*High calorific value:* Acacia, tagasaste, casuarina, yellow box, red gum, etc. *Animal fodder and fuel wood:* Kurrajong, acacia and brachychiton. *Self-pruning:* Eucalypts, ironwood, bloodwood, red gum.	Choose trees that are coppicing or self-pruning. They are lopped for firewood when they are 4–10 cm in diameter, which is about the right size for cooking and heating stoves. They are harvested from Year Two with one-seventh of a whole fuel-wood forest being harvested every year. All the listed species are Australian examples of trees having high calorific value (they generate a lot of heat and so you need less wood). If you do not know your local firewood species ask your local timber mill.
Polewood	*Very durable poles:* These are used for fences, housing and furniture construction. Species include chestnut, raspberry jam acacia, black locust, cedars, some eucalypts, turpentine and red river gum. *Less durable poles:* These are used for scaffolding, formwork, chipping, fuel bricks, fibre, cellulose, stockfeed, and oils. Species include poplars, willows, acacia and Chinese elm.	These species are among the faster-growing and early-maturing trees. They are grown as a 'thinning' crop amongst slow-growing, high-value trees.
Long-term fine timbers	See Table 16.4	These are grown as an investment and add to the capital value of the farm from the first year of planting. Complementary pioneer, short-term species are inter-planted for their yields, such as leguminous trees and small cedars. They are used to make fine furniture, inlays, panelling and plywoods. Their harvest time is from 20 to 400 or 700 or 1000 years depending on the species and what you would like to leave future generations.
Special purpose forests	Many of these species are being lost and not replanted: • rattan palm • species grown for tree oils and traditional medicines • some bamboo species • species grown for their bark, such as cinnamon, quinine and quassia • basketry species • special timbers for tools, musical instruments, etc.	On 'spare' land plant forests that produce a special product or occupy a special niche. Look for the following places on your land which can yield tree products in danger of being lost as professional forestry turns to species monoculture: hedgerows, contour banks, steep slopes, roadsides, swamps, watersheds, acid uplands or alkaline/saline soils. You may decide to grow the rare and endangered species from special ecosystems.

Forests in other cultures and places

Forests are in trouble around the world. My friends Michel and Lynne Le Rouet in Normandy, France, where Michel is a fourth-generation cabinetmaker, tell me that until recently Normandy farmers at the birth of a child planted sufficient trees to provide for the house and furniture for that child upon marriage. Now, because trees are so few, Michel looks for oak trees in farm fields and buys them, one by one, from the farmers. He then carefully fells the tree and ages the timber before using it.

Despite the comforting sound of 'community

TABLE 16.4: LONG-TERM FINE WOOD SPECIES

Temperate	Transition	Tropical
Black walnut	Rosewood	Red bean
White beech	Cudgeria	Rain tree
Paulownia	Black ebony	Spanish mahogany
Blackwood	Jacaranda	Albizia
Tulipwood	Blackbean	Transvaal teak
Oaks	Burmese rosewood	Swamp cypress
Redwood	Teak	
Silky oak	Honduras mahogany	
Mulberry	Camphor laurel	
White cedar		

forestry', in India many farmers have found that because they did not have recognised ownership of the trees they planted, at maturity a landowner or even the local government officers came, felled the trees and sold the wood. The priority for community forests is to establish local control and management. Harvesting must always be for community needs and not for business interests.

In Vietnam, I found one village suffering because a non-government organisation (NGO) had planted all the hillsides, which previously had tertiary regrowth forest, to a eucalypt plantation. The people who used to harvest small twigs for fuel to cook rice or small-scale building were forbidden access and fined if found on the land. This 'community forest' contributed to the farmers' poverty and difficulties. However, here and there are signs of hope, places where forests are still considered valuable parts of the local ecosystem.

 Try these:

1. Draw up a list of timber species you want to grow and their products. Research their time to maturity, their adult size and their needs.
2. Add a harvest forest to your site design. Design a windbreak that will protect it. If you are in a city, then think where you would grow an urban forest.
3. Every year plant trees for future generations.

CHAPTER 17

Natural conservation forests: Zone V

Natural conservation forests are indigenous, old-growth or regenerated forests. As your Zone I garden is your food security, so these forests are the healthy backbone of your landscape. They have evolved over millions of years, under all the pressures of natural selection, so that finally what we see is a highly refined, complex, beautifully honed to survive on the soils and in the topography and climate where you live.

Forests are balanced so that they will be perfectly self-sustaining if left undisturbed. When the environment changes, then, given time, forests change. They are miracles. This is how eucalypts and acacias became Australia's signature trees—in response to a general drying of the continent. Trees are the signature of a place.

Everywhere, natural forests are being rapidly removed and remaining forests can have weed and feral animal infestations. It is almost impossible to replace natural bushland exactly because species we do not even know about may have disappeared. People quickly forget when forests are felled and regard the subsequent impoverished landscape as natural.

Although some non-timber products (NTPs) can be harvested, the trees should never be felled. These natural forests must be inalienable.

 ## Our ethical task is to:

- reserve, extend and care for all remaining forests
- establish wildlife corridors to connect them up.

 ## Our design aims for forests are to:

- preserve all fragments of natural bushland along rivers, waterways and farm boundaries, and cover all ridges and slopes greater than 18 degrees
- assess the health of a forest and any environmental threats to it
- establish natural conservation forests on degenerated land
- encourage wild native animals and plant species
- control animal and plant pests
- implement bush regeneration strategies.

 ## If we don't have design aims:

- forests quickly and quietly disappear
- springs, creeks and streams become ephemeral and quickly dry up
- future generations will be deprived of timber and NTPs and inherit greater instability of soil and climate
- other species will suffer extinction— absolute or local—from loss of habitat.

CHARACTERISTICS	APPLICATIONS
ENDEMIC FORESTS ALONG ROADS, VALLEYS AND RIDGES	• PROTECTS WATER AND SOIL FROM MANY FORMS OF EROSION • HELPS TO PROTECT OTHER ZONES
SITE FOR HIGH DAMS	• WATERS KEPT CLEAN AND SACRED
REFUGE, SANCTUARY AND SENSE OF PLACE, IS CONSTANTLY EVOLVING	• PLANT AND ANIMAL GENE STORE AGAINST DISASTERS EG FIRES, LOGGING AND FLOODS
FLOWS INTO ZONE IV, NATIONAL PARKS, STATE FORESTS AND CROWN LAND	• ACTS AS A WILDLIFE CORRIDOR
NOT CUT	• MAY BE HARVESTED FOR SURPLUS SEED/FRUITS
GRAZED AND BROWSED BY NATIVE ANIMALS	• NEEDS OCCASIONAL WEEDING OR PEST ANIMAL CONTROL
VALUE - PRICELESS	• EVOLVED OVER THOUSANDS OF YEARS • IS THE BACKBONE OF THE SITE

Figure 17.1 Characteristics of Zone V.

Ecological functions of natural forests

Like the food and harvest forests, this assembly of trees and their organisms:

- provides shelter and protection, and maintains stability of air, soil and water
- preserves the perfectly adapted genetic material of plants and animals
- buffers cyclones and tsunamis
- absorbs carbon dioxide
- if large enough, will offer sanctuary to indigenous mammals, birds and reptiles. In the event of fire or drought, cyclones or tsunamis, animals can move around this zone to escape the worst of the disaster, especially if it is linked to wildlife corridors.

A special benefit of these forests is that of conveying a sense of place for each region. When you think of northern hemisphere forests you picture deciduous or coniferous forests; in Australia you think of eucalypt forests; equatorial belts have rainforests.

Remnant species from previous continents and cultures still persist in uncut forests and many species belong to stock strains that have been extinct for centuries elsewhere. For example, the Sahara Desert has only five desert species left in some places and Australia is providing replacement stock. Australia's Wollemi pine, so ancient it is likened to a living dinosaur, would have been woodchipped if it had not been in an area of forest preserved as a national park.

Siting and size of Zone V

Zone V is usually contiguous with Zone IV and may merge through windbreaks and wildlife corridors with Zone III or even Zone II. Certainly it should cover all difficult sites, such as swampy land and difficult soils, and landforms with their own

TABLE 17.1: PLACES FOR ZONE V FORESTS

Areas	Position
Suburbs	Every suburban home can have a small Zone V which is not a forest yet preserves local species, especially those that are rare or endangered. This zone is best situated along the fenceline and species are specially chosen for their adult height to be appropriate for domestic-scale buildings (see Figure 17.3). Preserving these rare species is significant and symbolic.
Towns and cities	There are many sites, like old rubbish dumps, derelict railway yards, along rivers and harbour foreshores, to establish natural forests. In rural towns forests encircling the town, while using grey water and sewerage water, will filter dust and modify extremes of wind. There are two good examples of this. One is at Alice Springs, in the centre of Australia, and another is the small coastal town of Maryborough in Victoria. Eventually these new forests can become part of the national parks network or form corridors adjoining them.
Rural areas	Plan to revegetate ridges, slopes, rivers and roadsides with natural forests in order to hold soil on land, prevent erosion, keep moisture in the soil, and modify climate extremes. Wildlife corridors along waterways offer security to wildlife, while reducing flood damage and absorbing and filtering fertilisers before they enter rivers.

ecosystems. Zone V follows all roads, rivers, boundaries and ridges. Because of its greatly reduced need for inputs and human maintenance, this zone is usually farthest from Zone 0. Figure 17.2 shows Rosie's farm; when you compare it to Figure 16.2 (see page 162) you can see how the natural forest was extended from the original remnant vegetation. Your design will be stronger and more sustainable if you design your Zone V first because it is the protective zone. The areas left for planting or enterprises will then lend themselves naturally to these activities and will be protected, leading to a sense of 'farming in clearings', which is a whole site permaculture design goal.

The size of Zone V is determined by the productivity and size of the land. Normally, natural forests are smaller on highly productive land and larger on fragile or marginal land. They are larger on big properties and smaller on small allotments.

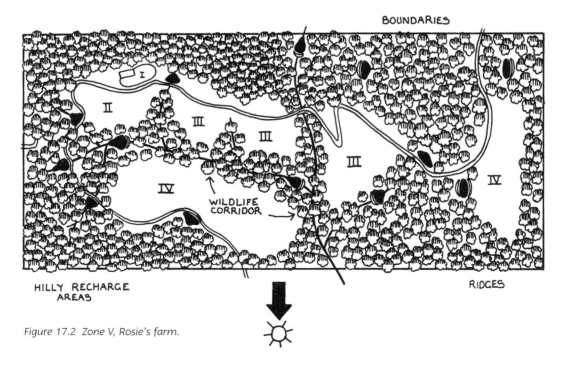

Figure 17.2 Zone V, Rosie's farm.

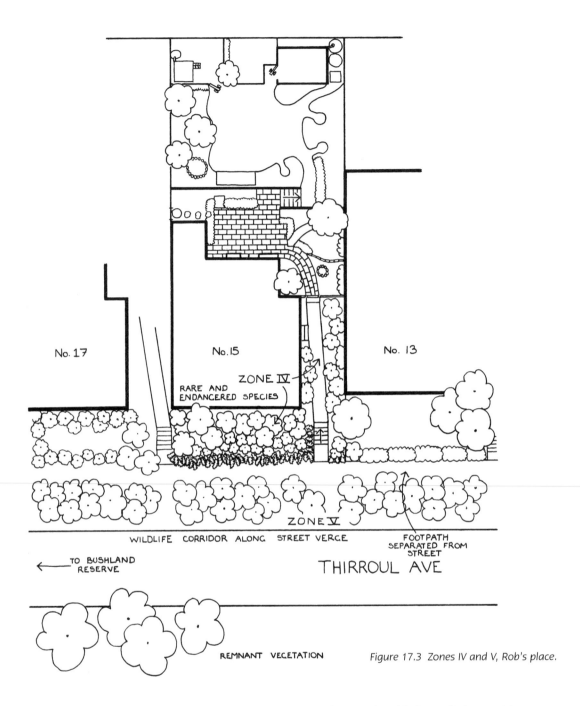

Figure 17.3 Zones IV and V, Rob's place.

You can compare Rosie's farm (Figure 17.2), where 40–50 per cent is dedicated to Zone V, with Rob's place (Figure 17.3), which has a narrow 3-metre strip at the front of his house for Zone V. At my place I dedicated 75 per cent to this zone because I live in the middle of forest and need to protect it from escape species. These are species that escape from domestic farms and gardens and become a problem for indigenous ecosystems through invasion.

Zone V forests should always link up with reserves, national parks and forests through corridors.

Reasons for forest degradation

Forests are used, abused and misused for a variety of reasons, most of which result from ignorance of what forests do for us and greed for timber

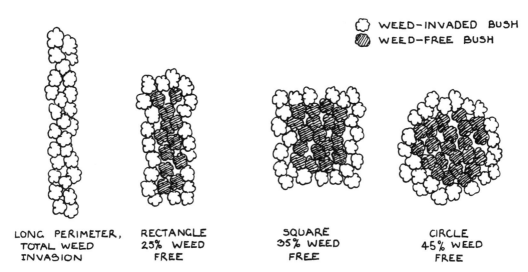

WEED-INVADED BUSH
WEED-FREE BUSH

LONG PERIMETER,
TOTAL WEED
INVASION

RECTANGLE
25% WEED
FREE

SQUARE
35% WEED
FREE

CIRCLE
45% WEED
FREE

Figure 17.4 Shape and weed infestation. Each diagram in this illustration represents a 2-hectare site. Although the area of each site is the same, the degree of weed infestation at each site varies according to the length of the perimeter.

products by big logging companies.

- The expansion of urban housing is accomplished most economically by the felling of every tree and shrub on the proposed development. The uncontrolled spread of housing absorbs both forests and valuable, fertile former farmland.
- Forests are destroyed in order to be replaced with industrial crops such as cotton, corn and soybeans, and pasture crops for cattle.
- The mechanisation of logging means that heavy machinery and round saws cut a big tree in 28 seconds.
- There is a vast hunger for paper produced from woodchips.
- Logging companies also making woodchips are primary polluters of rivers and streams from their processes.

Reclaiming and restoring remnant forests

Success in reclaiming and keeping remnant forests in good condition depends to some degree on the size and shape of the land; for example, a 2-hectare reserve will have about 0.7 hectares, or 33 per cent, undisturbed. A 64-hectare reserve will have about 54.8 hectares, or 86 per cent, undisturbed. The reserve should be as close to a circle in shape and as symmetrical as possible, in order to minimise the amount of edge vulnerable to invasion by animal pests and weeds (see Figure 17.4).

Weed and pest infestations

Whatever the causes of forest degradation, weed and animal infestations will always have to be

TABLE 17.2: WEED DENSITY

Class	Weed Density	Strategy
1.	Weedfree	Monitor to prevent weed establishment
2.	Few weeds	Minimum Disturbance Technique (MDT)
3.	50:50	MDT and spot weeding
4.	Dense weeds	Burn, slash, exclude light, graze
5.	Massive weed infestation	Remove with machinery, animal 'tractors'

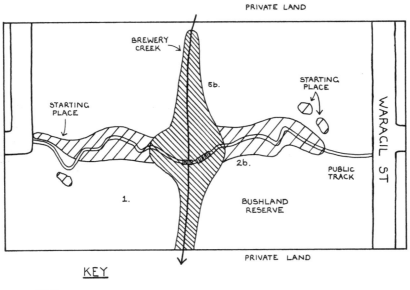

PRIVATE LAND

BREWERY CREEK

STARTING PLACE

5b.

STARTING PLACE

WARAGIL ST

PUBLIC TRACK

2b.

1.

BUSHLAND RESERVE

PRIVATE LAND

KEY

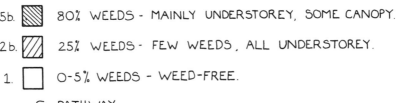

5b. 80% WEEDS - MAINLY UNDERSTOREY, SOME CANOPY.

2b. 25% WEEDS - FEW WEEDS, ALL UNDERSTOREY.

1. 0-5% WEEDS - WEED-FREE.

PATHWAY

CREEK

Figure 17.5 Weed analysis of Waragil Street reserve.

managed. There are several strategies that you can use to eradicate or manage them and the most ecological for weeds is based on the work of bushland regenerators in Australia, whose techniques have been very successful. When you have a weed problem it is enormously tempting to seize some tools and attack and clear the area completely. However, hold back. You will achieve better results if you follow these steps:

- Firstly, fence off the area from all animals likely to eat new regrowth. In many cases these are goats, horses and rabbits.
- Next, do a weed analysis.

Weed analysis and mapping

Decide the extent of surface area and mark the boundary on a map (see Figure 17.5 for an example using Waragil Street Reserve). Estimate in which structural layer the weeds are dominant. These are the structural layers of infestation:

- groundcovers and creepers, such as *Tradescantia* spp.

- intermingled with native shrubs, such as mickey mouse plant
- overstorey or canopy—camphor laurel, privet
- understorey—*Lantana* spp.

There are five grades of density of infestation. Table 17.2 shows the strategy most appropriate for managing each grade.

Cause, location and action required

Examine the conditions the weeds are growing under. You can often change the physical conditions to kill or weaken weeds. Check whether you need to make the soil wetter, or drier, or even change the pH. For many weeds, shading is very effective. Table 17.3 is an analysis with suggestions for actions to alter the environmental conditions.

Study the column listing the causes of infestation because the cause will give you the clue to controlling them. To begin with, simply change the growing conditions or select an appropriate method of removal based on how and when the weed propagates.

TABLE 17.3: WEED INFESTATION

Causes of infestation	Originates	Action
Overused, compacted soil	Under single trees, along fencelines, pathways, around watering places	Reduce stocking rates, change pathways, remove traffic
Incompatible use	Horses, 4WDs, trail bikes, etc.	Fence off, exclude from land, stick to trails
Increased volume of water	Houses, factories, shed roofs, dam spillways, overwatering	Decrease water use, recycle, redirect to orchards, forests
Erosion	Wind, water, land fatigue, compaction	Windbreaks, trees on slopes, terrace, change cultivation methods
Water pollution	Fertiliser, industry	No use of chemicals, reduce herbicides use, barrier plants, filters
Air pollution	Vehicle emissions, factories, detergents	Hardy plants, get rid of car, filter plants, refuse to use pollutants, change driving behaviour
Rubbish	Garden dumpings	Ask to stop, remove
Fire	Hazard reduction burning and prescribed burns	Stop burning
Feral animals	Disperse seed, kill native species	Eradicate, reduce numbers, fence out
Fewer native animals	Fewer pollinators and predators and less seed dispersal	Rebuild habitat, exclude feral species.

If the weeds are annuals near seeding, slash at the budding stage (before flowering) to prevent seed dispersal. Remove any seedheads. Slash biennials annually, before flowering. And for perennials, pull all seedlings at knee height after rain.

When these methods are insufficient, try mowing and burning. Your final resort is to mechanical clearing and herbicides.

The Minimum Disturbance Technique (MDT)
This strategy was pioneered by sisters Joan and Eileen Bradley in Sydney. Their aim was to restore and maintain ecosystems so natural regeneration could take place. It was so successful in restoring forests and bushland that it has now been extended to include reintroducing rare and endangered species into localities where they were known to have existed previously. Bush regeneration is now a career option and is taught in TAFEs and universities. It is most successful:
- where native plants can colonise the site by seed or vegetable means
- in areas sensitive to erosion
- in areas likely to be overused.

Its tools are trowel, pliers, secateurs, bowsaw and tree-loppers and it relies on the following principles:
- Work from the least infested area to the most infested area.
- Minimise soil disturbance because disturbed soil once exposed to the light usually germinates immense quantities of weed seed.
- Allow indigenous plant regeneration to dictate the rate of weed removal.
- Going back over initial weeding is crucial.

To practise MDT:
- Mark off an area as near as possible to a circle, and so define a small boundary for reinfestation. A long line is much more open to damage and reinfestation.
- Start where the vegetation is the least infested with weeds and work inwards, taking care not to disturb the soil.
- Allow natural reseeding and growth of new seedlings to take place before continuing. Re-weed the first section before continuing to a new area.

● Spot-weed around groups or individual indigenous plants and gradually let these weed-free 'spots' join up as natural seedlings emerge.

Establishing natural forests on cleared land or rural land

Dieback, like soil salinity, is such a serious and wide-ranging phenomenon of rural land that it is worthwhile revising its causes:

● grain production and grazing
● high nitrogen levels
● a proliferation of scarab beetles on pasture
● a build-up of tree-skeletonising pests
● a lack of understorey like *Bursaria* spp.
● the loss of bird and insect predators
● cool-fire burns leading to an increase in *Phytophthera cinnamomum* and the removal of major tree cover.

It is impossible to establish a 'natural' forest on cleared land or land affected by dieback since we never know exactly what was there in the first place, nor the proportional mix of species. The very act of clearing forests changes soils and microclimates to those less favourable for their survival and usually makes replanting difficult. However, there are a number of strategies to establish a basic native forest structure.

Strategies for bare ground along rivers, hills, roads and farmland

● Start with indigenous pioneers.
● Fence off an area of cleared land and remove weeds by slashing or mulching or by using animals as 'tractors'. When rehabilitating a river or creek, fence off 30 metres on both sides of the streamline (see Figure 17.6).
● Apply controls for pests such as rabbits that can destroy new plants (see Chapter 21 for strategies on pest management).
● A 600-metre-thick edge filters the wind of dust, wild seed, viruses and insect pests and inhibits the entry of feral animals. It also

Figure 17.6 Fencing along watercourses. A minimum area of 30 metres on each side of the watercourse should be revegetated.

cools or warms winds to reduce both wind speed and temperature extremes.
● One old tree, when fenced off, can throw sufficient seed to generate a copse of fairly dense trees (see Figure 17.7).
● Encourage birds to deposit seeds via faeces by putting stakes in the ground to act as perches.
● Weed carefully as plants start to grow.

Conservation zones will *always* need management. Keep watch for signs that the land is regenerating. In Australia, echidnas are a sign of a healthy habitat and bushland.

Mimic nature when planting

Establishing your indigenous forest will be more successful if you mimic nature. Instead of planting climax species, begin with dense plantings of pioneer plants, usually the nitrogen-fixing species which emerge first after land clearing or fire.

When the seedlings are 18 months to two years old, make pathways through them and plant climax species seedlings. The climax species will have better survival rates because the pioneers provide shade, a windbreak, humidity and improved soils. The establishment of the whole forest is faster and more successful this way. Even losses from insects and other pests are reduced.

Use one or a combination of these techniques to establish new forest:

● Hand broadcast a seed mixture of known indigenous pioneer species.

- Apply topsoil with a known seedbank from neighbouring land.
- Reseed from one tree—this is the most effective method of all (see Figure 17.7).
- Broadcast seed mix from the air—an especially good method for steep land.
- Scatter branches of desirable species bearing ripe seedpods which will shatter and drop seed over bare soil. This is very effective as the leaves act as mulch and a seedbank is restored.
- Broadcast pelleted seed by machine or hand. A camel pitter is a light implement drawn behind a utility (or pick-up), which scoops out a cupful of earth and drops seed into the hollows left behind. Dust and organic matter will gather in the hole, and when it rains and conditions are right, the seeds will germinate. The camel pitter's action is similar to that of an animal's hoof on the ground, and the implement has been used with good success in desert regions of Australia.
- Spiral-plant with a camel pitter in desert areas, where seedlings quickly dry out. Drill in large spirals for the seedling rows. The trees protect each other, the humidity is higher, the soil is held together and water is used more efficiently.
- Hand-transplanting seedlings in tube stock or pots is very laborious, so for large areas use machinery. There are two main ways to do this. In the first method, a tractor pulls a cart carrying people who drop the seedlings into prepared holes. In the second method, the land is ripped along its contours, the furrows are filled with water and then a vehicle drops off tube stock while people walk behind planting the seedlings. One person can plant up to 600 trees a day. Landcare groups use this technique to establish large windbreaks and shelterbelts and as many as 20 people will help to plant on one person's land, who will in turn have their own land planted.

Once your forest has been established, it must be cared for. This means keeping it from being eaten and excluding weed species. So fence it off from animals until the trees are mature—from three to six years old. The most pleasant way to protect your forest from weed invasion is to walk through it regularly and pull out the invaders.

 ## Try these:

1. Place Zone V on your site plan.
2. Find some weed-infested remnant vegetation and do a weed analysis covering:
   - weed density
   - structural infestation—that is, whether it's understorey, canopy, etc.
   - causes of weeds and possible control strategies.
3. Find a place where you can plant a small forest. For example, your church, pagoda or local school. Farmers may be pleased for you to plant a forest on their land.
4. Assess the land:
   - Are you starting from bare ground or weeds?
   - Research local climax and pioneer species.
   - Make a plan for the best season for planting.
   - Make sure plants will be safe from hungry animals.
   - What will be your establishment techniques?

PREVAILING WIND

Figure 17.7 Natural regeneration can be achieved by fencing the leeward side of a tree. The seeds will be blown into the fenced area and the seedlings will be protected from grazing animals.

CHAPTER 18

Other cultures and places

World biozones are those broad areas of the world where the different climatic regimes have given rise to different plants, soils, animals and farm practices. Within each there are many, many examples of sustainable societies where people live well and have integrated their activities into natural ecosystems without destroying them. These societies are intrinsically fascinating and visitors are drawn to them; often they are eco-tourism destinations.

Culture and food production are linked. F.H. King in *Farmers of Forty Centuries* explains how Chinese, Japanese and Korean farmers maintained highly productive 'permaculture' systems over thousands of years without degrading the land. In their fascinating book *The Gobi Desert*, Mildred Cable and Francesca French describe the care and husbandry techniques of oasis dwellers and nomads of the Gobi Desert. These two books are inspirational reading because they show what can be achieved under very difficult conditions.

In recent years many garden and farm cultures have collapsed as a result of the imposition of technologies required for cash crops. Indigenous peoples, forced to abandon their role as carers and maintainers of traditional systems, drift to towns and cities and live as landless fringe dwellers.

Permaculture principles apply to all climates and zones, not only temperate ones. But permaculture strategies and techniques are specific to different sites and regions and unthinkingly transferred strategies can be environmentally destructive. In this chapter we look at the three major biozones and refer to some smaller ones.

Our ethical task is to:

- respect and understand the specifically evolved living and cultivation strategies and techniques of cultures different from our own.

Our design aims for working in other cultures are to:

- discover and respect traditional techniques before we even contemplate transplanting foreign industrial technologies and strategies
- work from principles to practices
- realise that the impact of new technologies will be greater and the inter-relationships more complex than we can imagine or predict
- start small and proceed slowly
- think locally in terms of expertise, materials, systems and species
- design food and water security systems for each biozone.

If we don't have design aims:

- desertification occurs.

Causes of desertification

When biozone differences are ignored and inappropriate strategies are applied, desertification occurs. Areas particularly vulnerable to this are productive marginal land and anywhere where production is declining. Some examples are outlined below.

Deserts of salted soils

Inappropriate strategies of cultivation using soluble fertilisers, excessive watering and high evaporation rates in naturally saline soils has turned trillions of hectares into deserts and salt lakes in arid areas of Australia, Pakistan and the United States. These strategies were directly imported from high-rainfall temperate areas.

Destruction of watersheds

Massive dams in tropical monsoon areas have contributed to unnaturally high evaporation rates. Watersheds below the dams are drying out as rivers are robbed of environmental flows. Dams in Nepal are causing rivers in Bangladesh to dry up. In addition, indigenous people are displaced so the dams can be built and valuable habitat is lost. The dams are used to water exotic cash crops often unsuited to the conditions.

Polluted watertables and loss of biodiversity

Industrial cropping techniques have resulted in polluted rivers and watertables, soil degradation and the loss of species. While these problems occur in all the biozones, the effects are worse in hot wet and hot dry biozones. Here the water evaporates quickly and the chemicals are concentrated in the soils to the point of toxicity. In hot wet biozones there is insufficient soil organic matter to absorb excess chemicals, and in dry areas the chemicals are not dissolved and absorbed because of lack of water and organic matter.

Inappropriate large-scale forestry

Many forestry enterprises use inappropriate species of trees which contribute to drying out the watersheds, the loss of genetic diversity and reduced productivity per hectare for local people. Village people in India are ripping out eucalypt seedlings in plantations because they want the return of local indigenous forests with their associated yields of herbs, barks, medicines, dyes and foodstuff. In Vietnam, local farmers were forbidden to harvest twigs from eucalyptus plantations where once they had harvested multiple yields from their forests.

Settlements in deltas and floodplains

The recent destruction wrought by Hurricane Katrina illustrates what happens when people try to manipulate tropical deltas by draining them, imposing bland infrastructure and trying to make them like temperate zone settlements. Traditionally, people living in deltas had strategies for dealing with massive floods; however, as settlements become more international and generic the people lack the skills and knowledge to cope. In Vietnam where there are three or four cyclones each year, it is the local inhabitants who come to the rescue and shore up the levee banks. It is their culture and tradition.

Sustainable living in different regions

Cultures are sustainable and societies are successful when the people live harmoniously within the limits and work with the advantages and characteristics of their biozone. Table 18.1 highlights the main characteristics and differences in major world biozones.

Cool wet landscapes

These have an attractive landscape pattern of small houses built of local materials grouped together into villages. Some examples can still be seen in Ireland, Norway, England, northern France and

TABLE 18.1: SUSTAINABLE LIVING IN DIFFERENT REGIONS

Factor	Tropical	Temperate	Arid
Soils	20–25% nutrients Little humus Leached fast No surface mulches	90–95% nutrients High organic matter Slow leaching Natural mulches	Plentiful clay minerals Low organic matter Often salty No mulches
Plants	Hold 78–80% of nutrients Massive biomass Stacking occurs Use tree crops	Hold 5–10% of nutrients Humus is vital Deep-rooted trees Deciduous species	Adapted for dryness Deep-rooted or ephemeral plants
Landform	Shaped by water Deltas are common Water-rounded shapes	Shaped by ice and water Angular and rounded	Eroded by wind and water Special shapes, buttes and mesas

Water effects	Aquaculture productive
	  and natural
		Surface storage abundant
		  and good
			Underground storage

Biomass	Continual growth
	Rainfall triggers germination
	  and flowering
		Seasonal growth
		Daylength and temperature
		  regulate growth
			Plants endure or escape
			  dryness
			Rainfall triggers periodic
			  growth

Cultivation	Machinery disastrous Claypans develop Soils quickly infertile	No tillage Grain crops Use mulches	Spot strategies Opportunistic—plant when it rains

Strategies	Nitrogen-fixing
	  groundcovers
	Nutrient cycling
	Stacking
		Use heat and light
		Add mulch
		Glasshouses and other
		  structures
			Drip irrigation
			Shade for young growth

Limits to growth	Soil poverty Heat Forest clearing Introduced crops	Temperature and light Fire and frost Chemicals	Dry periods Temperature extremes Overwatering Monocultures

Germany. Villagers have their own food gardens but share commons of grassland, permanent forest and water, which protects cultivated areas, rivers and streams and gives everyone access to essentials. Where villages are situated in hilly country the farming has developed a practical and recognisable profile for solar gain in houses and fields. This design can be adapted by individual farmers in this biozone (see Figure 18.1).

These lands were shaped by ice and so toothed mountains and steep slopes are part of their landscape. Rich river valleys exist and are farmed.

Hot wet biozones

These have historically supported millions of people in the deltas of the Ganges, Irrawaddy, Red, Mekong and Indus rivers. The farming is highly productive, self-sustaining and non-polluting. As in the cool wet climates, village houses, built of local materials are grouped, and each has its own food garden. Protection for the village is from orchard trees, bamboo thickets and trees for timber and non-timber uses. Legumes are planted on the levee banks of the terraced rice paddies, and ducks, geese, frogs and fish fertilise the rice fields and control the pests (see Figure 18.2).

Many farms still follow this traditional pattern; however, where governments or companies remove the levee banks and the fruit and forest trees to increase the total area available for chemical hybrid rice-growing for cash or for export, the fish die in the rivers and canals. As a consequence farmers

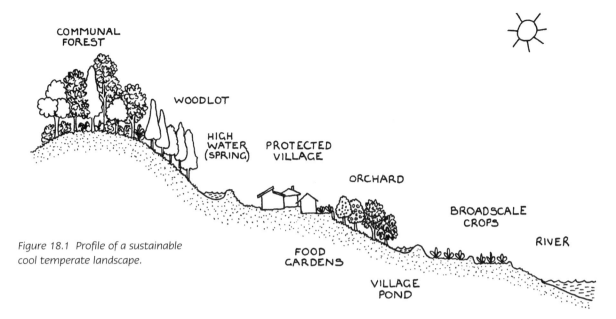

Figure 18.1 Profile of a sustainable cool temperate landscape.

either go into debt to buy capital-intensive machinery (and crop prices do not rise to cover these costs) or are dispossessed of their ancestral land and move to towns.

The landscape of hot wet zones is rounded by water erosion and this shape determines how soils, water and fires react, and limits the design. Land between hot wet and arid biozones can quickly become deserts if forests are removed. These forests create thermal zones.

If you look from the top of hills and mountains in humid zones, you see areas of upper slopes where soils will be unstable if the angle is greater than 18 degrees. Forests need to be inalienable for other uses. Water harvested here is clean and can

easily be directed by gravity to points lower down and used for power. Key points are critical for water control and diverting water along contours. Cultivation is kept below the key points. Housing is well placed here. Grey water is placed below the housing and key points. In rice cultures, very steep slopes are terraced.

On flat valleys mulches are used on cultivated areas and rivers and creeks are protected. Swales intercede run-off. The lower slopes are for mixed cultivation areas and fire control.

Hot dry biozones

These biozones are very complex. Most natural deserts and arid zones were sensitively managed by

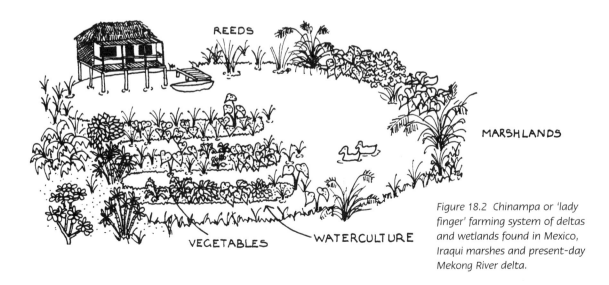

Figure 18.2 Chinampa or 'lady finger' farming system of deltas and wetlands found in Mexico, Iraqui marshes and present-day Mekong River delta.

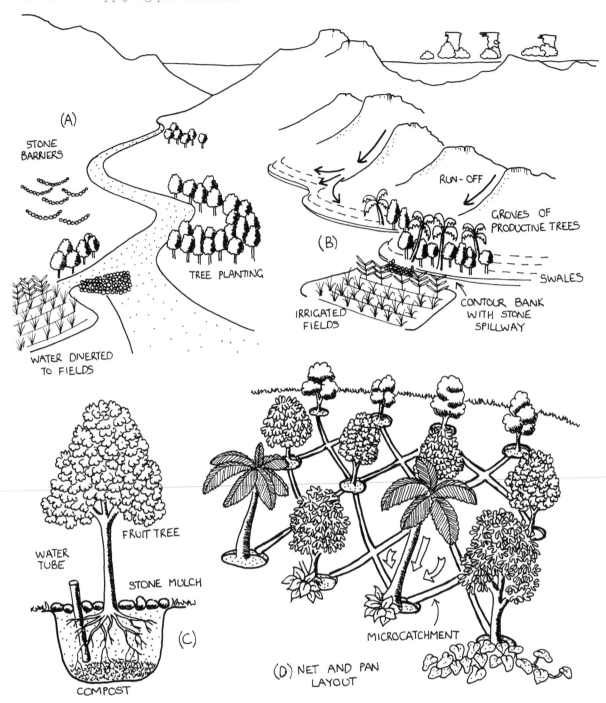

Figure 18.3 Sustainable strategies for drylands, stone barriers and trees planted on riverbanks
(A) check seasonal floodwater velocity, trap silt, reduce erosion and allow groundwater reserves to be recharged.
Water can also be diverted, allowing farmers to make use of seasonal flows for field preparation, planting and
production. In smaller catchments (B) silt and nutrient-rich run-off is intercepted by swales. Contour banks or 'bunds'
with stone spillways collect excess run-off and allow controlled irrigation of fields. Making efficient use of minimal
precipitation is vital in dryland areas; mulching (C) and creating microcatchments (D) are other useful strategies.

people who travelled around their lands as they gathered food and tended their animals and precious water supplies. Arabs, Tibetans, Bushmen and Aborigines all had extensive and precise

environmental knowledge which enabled them to live well in areas which most people today would consider uninhabitable without many inputs. A brilliant example is the Bishnoi people of the

Rajasthan desert in India whose annual rainfall is a mere 1 centimetre per year yet who live well and are healthy because their society maintains strict ecological laws.

Water-conserving strategies, strict controls on consumption and the conservation of valuable species are the principle reasons these dry-land farmers have succeeded. Farmers in low-rainfall areas use special techniques to catch moisture through condensation, such as using hard 'run-off' areas and selecting storage places as 'run-on' areas. Some Middle Eastern farmers dig underground tunnels to avoid the high evaporation of the deserts. Figure 18.3 shows mountainous dry land at (A) and flat and arid land at (D).

Settlements should be placed on the lower slopes away from the hot sun sector. These slopes are cooler, sheltered from hot winds and less prone to evaporation. At this point, run-off water from the slopes can be captured and the soil too will be finer and richer. There is often a water lens or aquifer at these points which can be tapped. Houses often have thick mud walls and are placed close together for coolness.

The agriculture is opportunistic and farmers wait and are ready for rain when it falls. Water is captured by land shaping with swales, meanders, and even on hills. In really dry areas around oases, camels and goats form the main diet with a variety of milk and cheese products. People move constantly to care for the water and grow dates, oranges and grains such as millet and sorghum.

Never, never pollute deserts, because they lack the micro-organisms and vegetation that perform the normal filtering and cleansing functions in other areas.

Other biozones

Smaller biozones exist within the larger ones, such as the mountain climates of India, Africa, South-East Asia and Central America. The villages are small, located on a ridge, and sited to gain maximum sun. Paths are usually along contours and forests are above the villages while the fields are below. On Bali, one village has never permitted its forests to be cut. Today this village still harvests valuable non-timber crops, doesn't suffer from floods and is far more prosperous that the neighbouring villages dependent on tourism.

In northern Vietnam, for centuries farmers practised an integrated small-scale agriculture that was highly productive. After the political and economic reforms of Doi Moi, the farmers moved from the collectives back to their traditional lands, and their practices were revived and gave them a good living. This form of traditional farming is still government supported and has been supplemented by permaculture teaching across all the northern provinces.

Vietnam and Mexico have chinampas, highly productive water systems in which the land reaches into the delta like fingers, resulting in a long perimeter and a variety of microclimates. Trellis crops are also grown over the water.

Other landscapes requiring special design strategies include coasts, high islands, low islands, wetlands, estuaries, cork-pork forest in Portugal and swidden areas (see Figure 18.5). The Boxing Day tsunami of 2004 showed that where strong coastal forests and windbreaks of, say, bamboo and mangroves remain, the damage sustained is much less than where these have been removed.

New sustainable cultures

Under the pressure of extreme consumerism and ecological breakdown, groups of people in each of these biozones are designing and building eco-villages. Some of these are sited on degraded farmland, some are close to older tribal cultures, and some are close to big cities. One of these is on Mindanao Island in the Philippines, where the dispossessed indigenous people—the T'Boli—have been assisted to buy back their ancestral lands and establish home food gardens or eco-farms. These 2.2-hectare farms are designed firstly to meet the basic needs of the T'Boli for food and shelter, and secondly to supply cash by selling excess produce to the local markets.

These eco-farms are very similar to permaculture zones in their functions and include food gardens, tropical orchards, alley cropping, animals for manure and work, and the manufacture of tools and fabrics with local materials.

There are now thousands of social and ecological experiments all over the world. Auroville in India has a pattern of small villages around a central core. Other communities work in shared land settlements.

In all of these, people are striving to change destructive patterns of land use and consumption. One factor common to the success of all these communities is the appropriateness of the design for the land. Australian suburbs are especially well placed to become eco-communities. An interesting example is the Penrose Permaculture Community in Australia, which was permitted a Multiple Occupancy title to their land because the local government office declared that the land was useless for conventional agriculture. The residents decided they would have only food gardens and allow the original vegetation to regenerate. After more than ten years the regrowth is extraordinarily beautiful and the indigenous animals fearlessly move around the residents because they have come to recognise the land as a sanctuary. The people live in simple mudbrick or recycled timber houses, with

Figure 18.4 Sustainable home VAC gardens of the Red River delta near Hanoi in Vietnam (VAC is an acronym for water, plant and animal ecosystem).

TIMBER TREES

PIG PEN

PIG FOOD

FISHPOND

NUTRITION PLOT

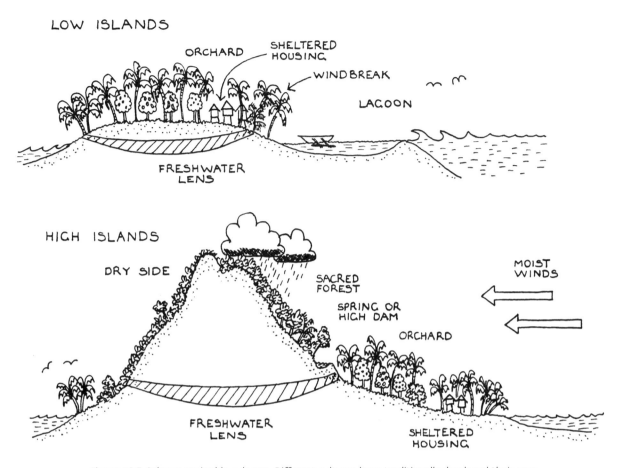

Figure 18.5 Other sustainable cultures. Different cultures have traditionally developed their own gardening and farming systems appropriate to their climate, landscapes and available food crops.

solar electricity and water tanks. Because their costs are low they do not need large incomes and can afford not to turn the land into a full-scale farm.

Originating from permaculture there is a worldwide movement of eco-villages known as GEN, the Global Ecovillage Network. And there are thousands of these villages, some in towns, others in suburbs and most in country areas close to towns. Some have succeeded for many years and are very settled. They are recording their successes and difficulties, creating a body of knowledge that may be important for all societies in the future.

Try these:

1. Choose a biozone that has always appealed to you. Describe its soils and vegetation. Now design a cultivated area which perfectly fits the natural features of this landscape.

2. Return to your plan and see whether you have really accounted for the special cultural and ecological features of your biozone. For example, I live in the mountains where the modern siting of houses on southern shaded slopes makes no sense because houses need heating 24 hours a day and even electric lights on in living areas. If my neighbourhood had been well designed then all the houses would have been along sun-facing slopes, with living rooms warmed by glass and lit by natural light. And it would look right in this landscape. Houses would be of wood and stone and protected from the cold winds by forests of natural vegetation left untouched, while food gardens and orchards would lie in front of the houses, down the slope.

CHAPTER 19

Permaculture at the office, shop and factory

Within 20 years more than 60 per cent of the world's population will be living and working in cities and towns. Offices, shops and factories are voracious consumers of non-renewable resources. They chew up energy, consume unsustainable quantities of water and spew out waste. They consume and waste more resources than households do.

The application of permaculture ethics and principles has a major impact on improving conservation practices and, at the same time, improving work environments, products and relationships.

Many workplaces are soul destroying or plainly unhealthy. Airconditioners leave pockets of stale air, building materials such as glues 'gas-off', and floorings and paints also emit allergenic or toxic materials.

As the number of people living in cities increases, we know that non-renewable resources are declining quickly. This is a time when demand for oil is rapidly overcoming supply and daily use is greater than daily extraction. As China, India and Brazil continue to demand fossil fuels the effect of this on work environments is an increase in the price of raw materials, transport, food and services.

Thoughtful employers and employees are aware they are now beginning to face a future of energy descent and climatic uncertainty. Many managers also know that immediate savings can be made by carrying out water, energy and product audits to see what current consumption levels are and then implementing plans to reduce them without compromising production or services. They also know that it is economical to adapt now while materials and transport are relatively cheap and there is time to change systems. Workplaces that adapt now have every chance of surviving future chaos. Remember the precautionary principle: we should take seriously and act on any serious or destructive diagnosis until it is proven erroneous.

We can all live comfortably in a society that uses about one-half the resources we use presently. Planning for a sustainable future is also good forward planning. Even if you're at the bottom of the office food chain, it's still possible to implement small changes and involve other staff members with whom you're friendly. Soon, these small steps will gain momentum and you might end up having a bigger effect on your workplace than you'd ever imagined.

Workplaces are really an extension of Zone 0. You can see that the permaculture ethics match a work environment well. Just to revise them:

- Care for people.
- Care for the Earth.
- Reduce consumption and distribute surplus.

To work in a just, co-operative and sustainable environment means that everyone agrees to audit consumption and value products and services. It removes the uncomfortable sense of acting destructively and contributes to pride and motivation among staff.

Our ethical task is to:

- produce essential and high-quality durable goods and services
- complement products and services with those of other businesses
- provide social environments for production and promote knowledge and skill sharing
- work for change where we can and where it counts.

Our design aims for work are to:

- work in a safe and healthy environment
- reduce non-renewable energy, water, paper resources
- support local people, businesses and their concerns
- minimise and manage waste
- value work.

If we don't have design aims:

- bills/expenses will be higher and continue to increase
- the work environment may be unpleasant and perhaps unhealthy, costly and disputatious
- we will add to the mountains of polluting non-degradable rubbish
- the workplace can be vulnerable to major breakdowns and failures.

Care for people

Work often provides staff with their main social environment because families are smaller and dispersed or social mobility hasn't allowed long-term relationships to develop. Work can provide the same ethical framework of trust and co-operation among staff that occurs in friendships and families,

and this translates to pride in goods and services. It also creates loyal and interactive customers.

At work, the principle of right livelihood is important for staff morale. This principle states that work should not produce goods or services that damage society. Many workers live with conflict in their lives between caring, mindful behaviour at home and more ruthless behaviour at work. A work environment that allows people to integrate their private and public lives and rewards ethical behaviour reaches its goals with less staff friction.

There are guidelines for achieving social work norms that value people. Look at the list below and see how your workplace measures up.

- Value and respect people for their diversity.
- Remember that work is something of value to yourself and others.
- Work together and treat each other well.
- Teach people to monitor and reduce their ecological footprint at work and at home.
- Support local business and suppliers.
- Encourage staff to bring their lunch to work and to picnic in the nearest park at lunchtime to clear their heads and lungs. Eating takeaways costs the environment twice as much energy (embedded energy) as food prepared at home.
- Celebrate relevant special occasions, such as achieving major conservation outcomes or staff achievements.
- Suggest meetings be held in sheltered places outside when possible and ensure that these places are quiet.

Care for the Earth

The workplace can model good practices that staff can take home and apply there. Workplaces have cost the Earth to build and maintain. By treating them badly, or by constructing them carelessly, contempt is conveyed for the resources used and the work put into them. So we should apply the same design aims for them as we did for our homes in Zone 0.

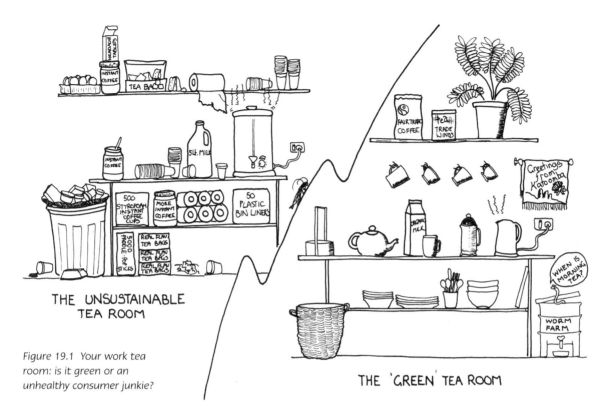

THE UNSUSTAINABLE
TEA ROOM

THE 'GREEN' TEA ROOM

Figure 19.1 Your work tea room: is it green or an unhealthy consumer junkie?

- Design or retrofit buildings to use renewable energy and efficient resources.
- Buy and use people-friendly, durable and safe equipment.

In developed countries there is often a lack of respect for buildings and equipment, which are treated as if they are distant from employees, and so damage or pilfering takes place because people have not understood how these actions damage everyone. When your buildings and equipment work well, morale improves, staff feel cared for, absenteeism decreases and productivity increases.

Request that your work canteen or kitchen uses crockery mugs instead of polystyrene and that organic waste is put into an office worm farm. Change your kitchen to resemble Figure 19.1.

It's also important to improve how workplace buildings function. First, involve the staff, explain why it's important and ask for their suggestions. Assign staff to do water and energy audits and report at staff meetings. Reward them with, say, a short article in your company's newsletter recognising their efforts. Let people know management takes the issue seriously. Ask staff to

identify how to get the biggest savings for the smallest investment. Publicise the savings made and what this amounts to; for example, the number of trees, or tonnes of carbon dioxide.

Your test will be to determine how well your workplace could continue to work if you had no electricity or water for three days. Ask staff to think about and plan for how the workplace would survive an emergency situation.

Reduce consumption and distribute surplus

You have designed systems from Zones 0 to V that are water and energy efficient, so you can now apply these to your workplace. Make sure systems are achievable and time oriented. Incorporate any creative suggestions from the staff, but ensure they meet the following criteria:

- Work with low-energy processes.
- Reduce and recycle waste.
- Monitor and reduce your workplace's ecological footprint in energy, water, waste, transport and food.

- Have a policy of giving, especially to local communities.

Strategies and techniques
Water conservation
- Do a whole site water audit and plan.
- Install low-flow taps, showerheads, toilets, washing machines and dishwashers.
- Install timers in showers and taps that shut off automatically.
- Use simple eco-friendly cleaning products, such as methylated spirits and bicarbonate of soda.
- Use only one or two eco-friendly washing products.
- Install rainwater tanks and use for toilets and washing machines.
- Plan to recycle grey water.
- Boil water, or filter it rather than use plastic bottles.

Energy efficiency
- Insulate ceilings, floors and walls.
- Weatherstrip doors and windows.
- Use low-energy light bulbs.
- Situate workstations to use natural daylight.
- Zone lights so they can be turned off when not in use.
- Keep the windows clean.
- Use desk lights, which use a fraction the energy of standard ceiling lights.
- Turn equipment off at power points when not in use.
- Use winter sun to warm buildings through glass.
- Ventilate rooms and allow cross-breezes in summer.
- Dress for the weather and not the airconditioner.

Office equipment and furniture
- Purchase equipment with low energy ratings.
- Use laptops instead of desktop computers.
- Use inkjet printers instead of laser printers.

- Use electric kettles instead of urns.
- Use photocopiers with low-energy standby and double-sided printing facilities.
- Minimise volatile organic compounds.
- Send all toxic materials to the nearest hazardous waste facility.
- Avoid chipboard, which gives off formaldehyde vapour. Use low-emission fibreboards.
- Small offices can use photocopiers in libraries and post offices.
- Use plant-based paints.
- Use hemp and cotton fabrics for upholstery.
- Choose recycled or plantation timber furniture.
- Select chairs built to last which incorporate recycled plastic or aluminium.
- Choose steel rather than aluminium.
- Switch from plastic packaging to minimal packaging and then preferably none.
- Ensure all computer parts, toner cartridges and photocopiers are returnable to suppliers who will recycle their equipment.

All that office paper
- Have only one printer and copier and keep them in a separate room so staff have to walk to them—consciousness raising!
- Have one wastepaper bin in another room—staff have to walk to it.
- Email or phone instead of sending faxes.
- Handwrite replies at the bottom of letters and memos.
- Keep mailing lists up to date to minimise wasting paper.
- Work on screen.
- Print drafts on used paper.
- Avoid using staples, plastic and wire binding.
- Ask suppliers to take packaging back.
- Reuse packaging such as envelopes and boxes.
- Purchase paper that is chlorine-free and made from post-consumer fibres.
- Send out-of-date stock to Reverse Garbage or give it away.

- Replace paper towels and tissues with reusable linen.
- Use unbleached or oxygen-bleached toilet paper.

Transport
- Walk or cycle to work.
- Car share or reward car-free days.
- Take public transport.
- Set low-emission and low-fuel vehicle consumption rates.
- Provide public transport access guides for visitors to your workplace.

Purchasing policies
- Employ and support local people and their products.
- Buy locally produced materials and ingredients. Choose organic materials first.
- Refuse to accept or use products and materials packaged in plastic.
- Ask your suppliers for their environmentally responsible purchasing program.
- Support a local Landcare group.
- Become involved with local environmental organisations and community events.

Recycling waste now pays. Most countries are running out of space for waste disposal and some send it offshore and pay poor countries to take it. Recycling companies pay for all metals and will even collect them. A company with a policy of recycling waste is now admired. The local community talks about and respects workplaces that recycle waste.

You will, of course, go much further than this outline. It is creative to take each of the permaculture principles and see what strategies and techniques you can design for your work. Work clean, work mindfully, work well and enjoy it!

Try these:
1. Ask a different worker each month to carry out energy, water and waste audits and discuss the results with co-workers. Encourage suggestions for reducing usage or recycling to some other good product or outcome.
2. Redesign your office, shop or factory so both the people and building work better.
3. Find out what your workplace's ecological footprint is now and then compare in a year's time.
4. What is the most likely disaster for your workplace? Make a plan to avoid, escape or endure it with the least distress.
5. What is your takeaway container count? Can you get it down to zero? Publish results on the work noticeboard.

Figure 19.2 Achieved: a healthy work environment.

PART FOUR

Adding resilience to design

If you implement a permaculture design alone, without co-operation from neighbours or watershed sharing, then your implemented design is vulnerable to many problems from outside your boundaries. Strength lies in co-ordinating design and implementation across bioregions and neighbourhoods. Storms, floods, droughts and pollution are much harder for one person to mitigate than many working together. Pest management and aquaculture systems are also much more effective if you work together.

In this Part, by anticipating problems that could destroy or weaken your farm or garden, you strengthen your design. Although you have designed and implemented excellent Zones 0 to V, your site is still susceptible to weeds, pests and catastrophes of various types. So, we will consider all these and also how to live well with wildlife.

CHAPTER 20
Design for disasters

Every day there are reports of disasters in all shapes and sizes occurring in every part of the world. In 2005 we saw the devastating effects of three natural catastrophes: the 2004 Boxing Day tsunami, Hurricane Katrina and the earthquake in Pakistan.

Although they occurred in different hemispheres and biozones, they shared the following characteristics with all disasters:

- they were situations which affected communities
- they had serious effects on those communities
- they demanded a total community response and action.

In addition to bioregional disasters we are now threatened with global disasters, such as global warming, petrol supply failure and freshwater decline. Meteorologists are trying to assess the impact of global warming on different climates of the world—this is difficult because long timeframes are needed to indicate trends. However, they say with certainty that weather will be much less predictable and violent storms, droughts and floods will be more frequent. These are forecasts we should heed.

In fast-onset disasters, it is the local people who save lives and property in the first 24 hours, before relief organisations can arrive with equipment. While the emergency services are organising vehicles, fuels, medicines, translators and other equipment, local people are rescuing people and reducing damage.

You can design landscapes that are resilient and which will evade or survive catastrophes. However, you need to know the types of disasters likely to

occur and what preparations you can make for them. In permaculture, as part of the site analysis, every site is assessed for its level of risk from a disaster, and a plan is developed for it. A site design incorporating all zones is usually not sufficient on its own; the best disaster planning and preparation is done by villages, neighbourhoods and communities. Design, planning and preparation are the keys to mitigation and survival.

 ## Our ethical task is to:

- save lives first and property last
- make comprehensive plans with our local community
- design to reduce the extent and impact of disasters
- design to endure or avoid the worst of the disaster.

 ## Our design aims for disasters are to:

- design landscapes and site buildings which can withstand or survive the disaster at its worst
- prepare for several types of catastrophe
- analyse the cause, frequency and duration of the disaster
- find out the frequency and impact of disasters

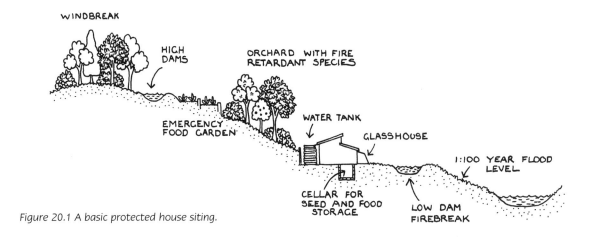

Figure 20.1 A basic protected house siting.

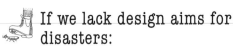

- apply the permaculture design methods of sector analysis and zones to reduce the risk of natural disasters from fire, flood, landslide and drought
- practise the precautionary principle, which states that if it is likely to happen it almost certainly will, so fix it or provide for that emergency now.

If we lack design aims for disasters:

- we are living with high risk and we and others are vulnerable to injury, suffering or worse
- repair or replacement of property, plants and animals can be impossible or very costly

- help may not be available when we require it (for example, government or emergency services).

Ecological functions of disasters

Although they are destructive and frightening, disasters often redistribute and recycle resources and nutrients. Floods bring valuable nutrient loads to delta and river flats. Droughts can control pest and disease populations. Erupting volcanoes spew out rich soil materials. Fires have a cleaning and regenerating role, and germinate indigenous seed in Australia.

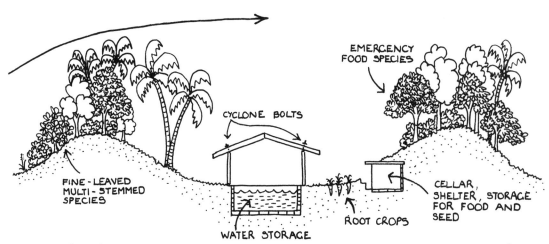

Figure 20.2 Siting for cyclone protection.

TABLE 20.1: POTENTIAL DISASTERS

Land-related disasters	Culture-related disasters	Secondary disasters
Fire	Global economics	Epidemics/pandemics
Floods	Nuclear accidents	Mass migration
Drought	War	Plagues
Cyclones, storms, hurricanes	Terrorism	Social collapse
Earthquakes	Climate change	
Tsunamis	Chemical spills	
Volcanic eruptions	Land degradation—deforestation	

Categories of disasters

Some disasters can be grouped because they are inter-related. For example, drought and war can lead to famine and plague. Others tend to be regional, such as earthquakes and volcanic activity. Some disasters may be more likely to occur where you live than others. For example, mountainous areas are susceptible to landslide once trees are cleared. While some disasters are related to geography and climate, others are more cultural and can be directly attributed to human activities. War, decline in water supplies, nuclear accidents and terrorism fall into this second category.

Table 20.1 is a list of potential disasters. The list has grown due to the increase in world population and more sophisticated technology. You may want to list industrialisation as a potential disaster because of its foreseeable consequences.

Making a disaster profile for your site

Drawing up a profile of the disaster ensures that all the major contingencies are covered. There is a great deal of information available about disasters. You can find weather reports, local newspaper articles, long-time residents' stories, and physical records along rivers or on trees. And of course you can search the web.

For your site make a profile guided by the list in Table 20.2. When the information is as complete as possible, design a plan for avoidance or endurance of the disasters.

Planning essentials

The first 24 hours after a disaster are crucial. Plans need to be made well in advance by the entire local community and everyone must commit to them. The design must consider possibilities for avoidance, endurance and/or escape. And you need to anticipate how long you will be in a state of emergency.

TABLE 20.2: DISASTER PROFILE

Disaster	Questions	Your answers
Cause	Natural or man-made?	
Frequency	How often does it occur?	
Duration	How long does it last?	
Speed of onset	What's the warning time?	
Scope of impact	Concentrated or large areas?	
Destructive potential	What is the population density?	
Predictability	Does it follow a pattern?	
Controllability	Are people helpless?	

TABLE 20.3: KEY CONSIDERATIONS

Structures	These should be made of materials and/or sited to withstand disaster. It is important to site a disaster refuge far from the likely centre, path or height of the disaster.
Escape routes over hills, etc.	Draw up and give the community maps of creeks, fire trails, back roads
Transport	How will people leave and where are the means kept? Foot, boat, bicycle, car, animal or other?
Evacuation plan	Who needs to be evacuated when, by whom, and transported to where?
Food security	Store seeds, food and water away from the likely centre of the disaster, such as in a cave, underground room, small mud house, on an island or in a dam. Ensure dry-food supplies are sufficient for the duration of the disaster and the number of people. Place an emergency/famine garden a long way from the disaster centre and grow simple hardy vegetables. Keep supplies of easy-to-grow seed stored in a dry place safe from weather and pests. Allow an absolute minimum of 3 litres of drinking water per person per day.
Cooking, heat and light	Ensure alternative energy supplies, such as firewood, batteries, solar panels, gas bottles, candles, matches, kerosene and lighters. For warmth, store blankets, hot water bottles, pullovers and coats, wind-, water- and fireproof clothes. Prepare an adequate store of cooking pots.
First aid and medicines	Nominate one or two people to be solely responsible for first aid. Equip them with a recommended first aid kit. Keep antiseptics and bandages as recommended by first aid groups. Plan for human waste disposal.
Tools	Make sure tents, spades and saws are on hand.
Communications	Fast, clear communication is essential. Ensure there are adequate (including back-up) whistles, flags, two-way radios and mobile phones. If all these collapse, designate one person as a runner who carries messages.

First and always, identify who is at risk in your community—especially children, and sick and elderly people—and either move them first out of the area or to a safe refuge to allow others to work on the disaster. Secondly, move out or free animals if possible.

All disasters have a lifetime and the community needs to know how long to expect to be in emergency planning.

Every community needs an emergency or famine garden. This is small and intensive, has hardy vegetables and is sited away from the most likely path of the disaster. The community must ensure a drinkable water supply and maintain a seedbox of local hardy, easy-to-grow plants that can be re-established immediately danger has passed.

Every person in a community is to belong to, watch out for, and account for ten others. One person in each group must have contact with one person in another group. People can then be counted, and if one is missing, the communication is fast. Anonymity and isolation are risky during disasters.

Lack of planning and preparedness is a major factor contributing to loss of life and property. Table 20.3 gives the key considerations for a community. It is not exhaustive because different disasters will require different preparations.

Detailed planning for specific disasters

Detailed planning has two parts:
- forward planning and implementation
- action when the disaster comes.

In the next few pages we will look at how to plan for climate-based disasters, earth movements, global disasters and fire.

Climate-based disasters

Floods, cyclones and droughts are not disasters where they occur regularly and people have adapted their life and food production to them. For example, the huge floods of the great deltas of the world such as the Mekong and the Ganges were seen as bringing renewal of soils and enrichment of floodplains. Cyclones occurring on coasts where bamboo and mangrove forests buffered them were simply part of a natural season, and housing and agriculture were adapted accordingly. Drought such as the six-month dry period in monsoon climates was valued by rice farmers who rested during these months. So these phenomena are disasters only when they occur more frequently, or when they occur newly in regions that don't have a history of them, or where the natural buffering ecosystems have been destroyed. (See Table 20.4.)

Floods

Floods are huge waters that overflow the banks of rivers. They are usually measured at 1-in-40, or 1-in-100-year frequency. They can be 'flash floods', which rise and subside over 24 hours; however, in some parts of the world such as southern Asia and South-East Asia, floods can last as long as three months. Many people die because early-warning systems are not in place and the floods arrive quickly. In areas where people are used to floods, one of the biggest dangers if a flood lasts a long time is the loss of children due to drowning because it is difficult to keep children confined during extended periods.

Floods are occurring more frequently because rivers are shallower due to increased sedimentation after tree removal, plus the loss of vegetation along rivers which buffered the energy of the floodwaters. In design, no permanent structures should be built within the 1-in-40-year frequency level.

Cyclones

Cyclones are massive spiral winds that develop over oceans and then move towards land, where they can tear down huge swathes of vegetation and houses and are accompanied by floods. They occur in the tropics during the hot, wet season. Traditionally, people in affected areas built appropriate housing, but with higher population density in areas where cyclones occur and increased climatic instability, the destructive impact can be much greater than forecast so that levee banks are insufficient and housing unprotected. Rebuilding houses without revegetating the area is simply a bandaid measure. If residents are to remain, they require the disaster planning discussed in this chapter.

Droughts

Australia declares droughts in areas where imported crop species and animals naturally adapted for wetter regions suffer water deficiency. There would be fewer droughts if crops and animals were suited to the climate and if people did not irrigate crops. However, there are seasons when the rains fail and where no provision has been made to store water or to reduce the dry periods. The site analysis and application of water harvesting and zones in permaculture enables most droughts to be endured or avoided if they occur naturally.

Earth movements

Seismographic data can give early warning of earth movements such as landslides, volcanic eruptions, tsunamis and earthquakes, but such warnings usually only occur shortly before the event itself. There are, however, regions of the world where land activity is more likely and where it is absolutely necessary to plan well in advance for these disasters.

Due to increasing population density, these disasters can have catastrophic effects on human

TABLE 20.4: CLIMATE-BASED DISASTER PLAN

Disaster	Action to be taken
Floods—last from days to three months	*Forward planning and implementation:* May last for up to three months. All structures below 1:100 floodline may be in danger. Plant densely along river banks to reduce flood energy and catch nutrients. Plant trees and shrubs on steep slopes to reduce run-off. *When disaster comes:* Don't enter floodwaters on foot. Don't drink floodwaters—they will be contaminated with sewage and chemicals. Don't use cars—wait for the water to recede or use boats. Carry clean drinking water or boil for 10 minutes minimum or use sterilising tablets. Watch children vigilantly. Many children drown during long floods. Ensure strict hygiene because epidemics usually follow floods.
Cyclones—usually last six weeks	*Forward planning and implementation:* Winds may last for six weeks and there may be a second cyclone. Build houses with cyclone bolts or palm leaves and open them up to let cyclonic winds pass through. Plant dense, flexible windbreaks, such as palms or bamboo, which absorb wind energy. Plant small-leafed and multi-stemmed shrubs with good root systems. Have a 'famine' garden in a very sheltered area. *When disaster comes:* Remain in your shelter during and after the passing of the 'eye'. Don't drink floodwaters—they will be contaminated with sewage and chemicals. Don't use cars—wait until waters recede or use boats. Carry clean drinking water or boil for 10 minutes minimum or use sterilising tablets.
Droughts—plan for years	*Forward planning and implementation:* Design drought-proofing strategies and techniques so people and ecosystems can survive through long periods. Water must be kept clean and not fouled. If drought is part of the normal weather cycle, then pastures and feed can be 'saved'. Do earthworks for water harvesting in droughts. Diversify your enterprises. Act as if a drought will follow the most recent rains. *When disaster comes:* Never lose supplies of seed or animals through greed or carelessness. Carry out drought-proofing and water-harvesting earthworks in dry times (see Chapter 4 on water and Chapter 6 on soils).

settlements if they are ill-prepared. For example, a major earthquake in Romania, where people were unprepared, caused a large number of deaths. In contrast, Japan has implemented planning at a national level, and all buildings are earthquake resistant and the people are trained to get under tables and stand in doorways. Although the country lies in a major earthquake zone and experiences reasonably frequent quakes, there have been no deaths and little damage to buildings for many years. (See Table 20.5.)

Earthquakes

These are sudden, usually short-lasting, earth movements that vary from mild trembles to large movements creating huge chasms in the earth. Seismologists can give warnings through their monitoring. There are earthquake belts around New Zealand, San Francisco and Japan which are monitored and warnings are given by radio and television according to a scale. People must be trained to react protectively and buildings need to be constructed so as to withstand the quake

TABLE 20.5: EARTH MOVEMENTS DISASTER PLAN

Disaster	Action to be taken
Earthquakes—over quickly but huge damage	*Forward planning and implementation:* Put in place fast communication systems. Move people permanently where possible. Teach community where first aid and tools to release people from rubble will be and where food stores are. Ensure safe water supplies. Ensure fire-fighting equipment is working and that people know how to use it. Teach people how to mark broken and probably live electricity wires. Construct buildings using appropriate principles and technology. *When disaster comes:* Move into shelters, under tables and doorways, and away from solid walls and structures. Local first aid, digging equipment and food supplies are urgently required to save lives in the first 24 hours. Shelters must be allocated for victims, children, women and elderly people.
Landslides—quick and serious	*Forward planning and implementation:* Revegetate slopes with permanent, never-to-be cut forest. Prepare another food site away from the likely site of disaster. Construct terraces where land may become unstable. Set up warning systems and monitor the site. *When disaster comes:* It is probably too late to save the lives of many in its path; however, other communities will need good organisation and tools, first aid, warmth, light and transport. Most lives have to be saved in the first 24 hours.
Tsunamis—fast onset, over in a few hours, but leave devastation	*Forward planning and implementation:* Many of the effects of tsunamis are the same as for cyclones—see Table 20.4. Bury seed in watertight containers for planting later on. Maintain root crop gardens with onions, garlic, carrots, potatoes and sweet potato, etc. Plant and protect buffer vegetation, such as bamboo forests, along ocean beaches, and reinstate lagoons. *When disaster comes:* If the first wave is not too large, move as quickly as possible away from the sea. Ensure emergency first aid and food supplies are available. Each person in a community must look out for and account for ten others. Use boats as soon as possible to conduct sea searches.
Volcanic eruptions—usually some activity before the eruption, which can be devastating	*Forward planning and implementation:* Set up local warning systems from seismic data, available for every household. Build emergency settlements where food and water are available. *When disaster comes:* At the first warning, evacuate children, the ill and elderly people to enable the subsequent orderly removal of animals and valued equipment.

effectively or, in the event of collapse, cause the minimum of damage.

Landslides

As forests are removed from steeper and steeper slopes, landslides are occurring more frequently. They give little or no warning and forward planning is the best way to deal with the threat. In Vietnam, Oxfam moved a whole village because of the likelihood it would be lost by a slowly moving hillside. On Bali, one village has kept its hills forested for more than 500 years and the people have always had food and wood for fuel. The neighbouring village cut down all its trees and experienced devastating annual floods and landslides. They are now reforesting their hills.

Tsunamis

These are giant tidal waves caused by undersea earthquakes. There is nothing that can be done to prevent a tidal wave occurring, but early-warning systems can give people enough time to evacuate. The best means of minimising or reducing damage are ecological, because naturally occurring coastal forests of mangroves and bamboo absorb and deflect floodwaters, while lagoons buffer floodwaters along the coast.

Volcanic eruptions

Volcanoes are openings in the Earth's crust through which molten rock, or lava, is expelled in response to pressure building up in the Earth's core. Volcanoes are classified as extinct, dormant or active. While seismologists are often able to warn of impending eruptions, there is probably little chance of escape if people or villages lie directly in the path of the lava flow. Despite this, many communities refuse to move off the slopes of active volcanoes and live lives of high risk. Once warnings are given, or even if people are feeling unsafe because of increased volcanic activity, they should move to known safe ground.

Global disasters

These are disasters which are very large in scale and which cross state and national boundaries.

Nuclear accidents

The biggest risk to most people is probably from leaks from equipment using nuclear technology, such as X-rays, irradiated food, planes and the photographic industry. Increasingly, nuclear weaponry, with its use of depleted uranium, is endangering lives wherever it is used. Most nuclear plants have fairly frequent small leaks.

Land degradation and famine

These have been fairly extensively covered throughout this book. Thoughtful development of a whole site by applying permaculture principles will strengthen most sites against disasters and the land will recover better.

TABLE 20.6: GLOBAL DISASTER PLAN

Disaster	Consequences	Strategy
Nuclear accident	All water will be radioactive.	Unpasteurised (raw) miso contains enzymes that help the body eliminate radioactivity. Some water-cleansing plants may be effective (see Chapter 4 on water). Vegetables grown in glasshouses will be relatively uncontaminated.
Land degradation/ famine	Stealing, raiding, trading of food supplies, illnesses, epidemics, and often many deaths will follow.	The application of sector and zone planning will alleviate problems and reduce risk. In particular: • implement Yeoman's Keyline to buffer drought and floods • have emergency food supplies. Set up hospitals and groups to handle deaths. Be ready to plant food crops as soon as practicable.
Epidemic/pandemic	Epidemics have a finite lifecycle, so wait it out.	Isolation is essential and breaks the cycle of infection, so stay home, but especially eat and drink at home (use your home garden and rainwater tanks). Calculate the amount of food and water required for the period and store it. Use non face-to-face communication systems, e.g. telephone and email. Help neighbourhood households with cooked food, medicines and other needs.

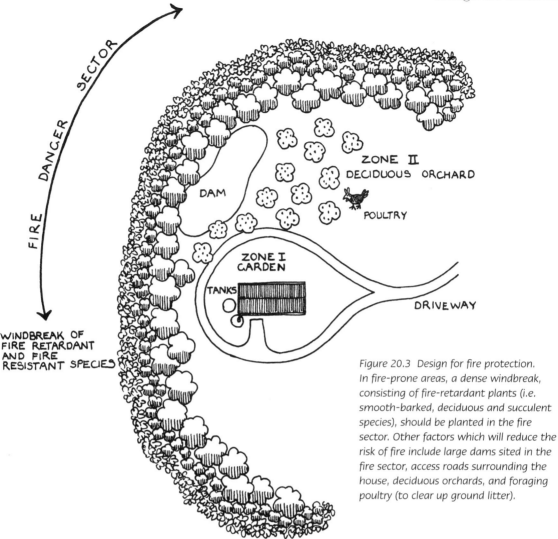

FIRE DANGER SECTOR

ZONE II
DECIDUOUS ORCHARD

DAM

POULTRY

ZONE I
GARDEN

TANKS

DRIVEWAY

WINDBREAK OF
FIRE RETARDANT
AND FIRE
RESISTANT SPECIES

*Figure 20.3 Design for fire protection.
In fire-prone areas, a dense windbreak,
consisting of fire-retardant plants (i.e.
smooth-barked, deciduous and succulent
species), should be planted in the fire
sector. Other factors which will reduce the
risk of fire include large dams sited in the
fire sector, access roads surrounding the
house, deciduous orchards, and foraging
poultry (to clear up ground litter).*

Epidemics and pandemics

These are far more likely to occur now with the huge and fast movements of people and products around the world. In an epidemic, the risk of contamination is through food and water but mainly people.

Fire

This is a major threat to many cities and towns in Australia and the United States, and with the advent of more frequent drought, fires will become more common everywhere. Eighty-five per cent of fires in Australia are lit by arsonists or escape from prescribed burns. It is not easy to understand arson or to know how to prevent it.

As with most other disasters, lack of preparation is the cause of most life and property loss. Fire survival is largely a case of being well prepared,

staying with your property and working with your community.

Remember that fires pass quickly and it is usually flying embers that start a blaze. There will often be good warning time. The best strategy is to always be prepared. (See Table 20.7.)

Factors traditionally considered to affect bushfire risk

- *Quantity of fuel:* A doubling of floor fuel produces a quadrupling of fire intensity.
- *Types of fuel:* Dry mulches of annual grass, cereal crops, and pasture burn very fast. Fibrous barks burn more than smooth bark. Pine needles burn faster than thicker matter.
- *Dry fuel and winds:* These increase the risk of fire.
- *Topography:* Fires move faster uphill.

Factors recently found to affect bushfire risk

Importantly, it now appears that it makes no difference how much bushland is around the house or residential precinct. What is important is what is happening when the fire is at its most intense: is the occupant remaining in the house and is the house protected from ember attack?

- *Poorly constructed structures* such as badly built sheds, timber fences and plastic gas lines are serious threats.
- *High fire intensity* is more risky than the presence of neighbouring bushland because of its live embers. This is known as ember attack.
- *Understorey bushland* should be kept because it slows down the wind and the fire.
- *Land clearing* speeds up winds and results in more intense fires.
- *Owners staying* with the house and knowing what to do reduces the chance of property loss and helps in controlling fires.

For further information on these factors, refer to *www.bushfire.com*

Forward planning in case of fire

Permaculture design by analysis and zone and sector implementation is extremely effective in minimising the risk of fire. However, there are areas of higher risk, such as Zones IV and V, which you may choose to sacrifice.

- In the sad case of losing all your vegetation you need to have built a small fire-safe core area, perhaps on the cool side of a hill or beside a dam or creek. Even a tiny garden of 3 x 3 metres will be sufficient to save your precious species.
- Surround this with an intermediate buffer zone with fire-safety features. These are, in fact, Zones I and II. Regeneration areas then surround these. Zone III is usually an effective firebreak, especially where it is planted with annual crops or fruit trees. Refer to Figure 20.1 and then look at Table 20.7 for the design strengths.
- Belong to a Community Fire Unit. You will be trained in saving life and property by the fire brigade. Leave bushfires to the bush fire brigades.

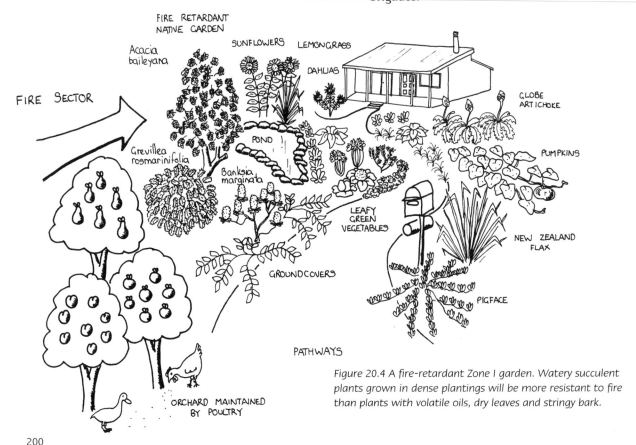

Figure 20.4 A fire-retardant Zone I garden. Watery succulent plants grown in dense plantings will be more resistant to fire than plants with volatile oils, dry leaves and stringy bark.

TABLE 20.7: BUSHFIRE PREPARATIONS FOR EACH ZONE

Zone	Preparation
Zone I garden	Around the house use damp mulches, irrigated crops and green mulches. Clean gutters, remove flammable items, and have buckets of water and mops ready to put out embers.
Zone II orchards	These are excellent firebreaks.
Water supplies	Water should be stored in irrigation/aquaculture ponds and tanks around the house, positioned between your home and the fire. Have supplies independent of authorities because these often fail. Plug and fill gutters, basins and baths. Move hoses inside.
Roads and paths	These should be circular or lead away from the direction from which fires come. Keep them clear.
Animal yards	Doors should open to the cool area. Consider animals in your escape plan.
Radiant heat barriers	Stone walls, mud walls, earth banks, concrete, bricks, thick low hedges, and white walls are all barriers to heat. Install flyscreens on the house.
Plants	Fire-retardant plants burn poorly—coprosma, wattles, agapanthus, willows, carob, mulberries, figs and deciduous fruit. Fire-resistant plants (eucalypts) will regenerate.
Fire shelter	Have somewhere that people can escape to if the house burns, preferably whitewashed and built of mud or rock, or a cave.

Disaster checklist

Chemical spills, terrorist acts, pandemics and natural disasters reach all corners of our globe. Revise the following basic disaster checklist and add anything else which is important for your situation:

- *Escape plan:* This accounts for everyone and prioritises who goes first.
- *Water:* You must ensure supplies of enough clean water to last the predicted duration of the disaster.
- *Food:* You will require enough dry food and staples to last the duration of the disaster. Remember to include some food that doesn't require cooking, as well as a cooking pot for when you can.
- *Medical care:* You will need a first-aid kit and trained people.
- *Clothes:* Remember to plan for wet or cold weather. Include a lightweight backpack and stout shoes.
- *Shelter and tools:* A tarpaulin, blankets, rope, tools, spades and axes are all essential.

- *Lighting and heating:* Ensure you have alternative sources of light and heat, such as candles and lighter or matches, torch, kerosene, lamps, gas and solar panels.
- *Communications:* Include a back-up such as a solar radio or phone.
- *Waste disposal:* Plan how to dispose of human manure safely.
- *Money:* Keep some hidden in a waterproof packet.
- *Documents:* Ensure the safety of any papers that identify you or are personally valuable or of sentimental value.

 Try these:

1. Estimate the biggest disaster risk to your place and draw up its profile.
2. Draw a detailed design for site protection.
3. Think of how your plan could be modified to take into account various scenarios. For example, how would you change the plan if there was a large reduction in the ozone layer?

CHAPTER 21

Managing pests: IPM

Integrated Pest Management (IPM), also known as Intelligent Pest Management, uses two or more strategies and many techniques to keep pest and disease problems at acceptable levels. Good gardeners are garden doctors who look at the plant, the soil and the amount of damage before they blame the pest. This chapter is about how to garden and farm well without pest infestations destroying your site.

IPM is based on skilled observation and deduction. In general, it takes about three years from the establishment of a garden, or from switching over to chemical-free techniques, to have a garden that has no serious pest and disease problems. Practise sitting in your garden and watching and understanding what is going on. Your garden is a dynamic insect refuge and zoo. Most insects are not harmful. They are an important part of the food chain and perform valuable work; they are pollinators, herbivores, carnivores and decomposers. Less than 0.1 per cent of the resident insects in your garden can be considered pests, and most of the rest are at work. When soil, water and plants are balanced, problems tend to self-correct and predators and parasites manage pests naturally.

Our ethical task is to:

- do minimal or no harm to the environment and any species because pests are part of the diversity of life; they only require controlling not eliminating
- apply deterrence rather than elimination strategies.

Our design aims for pest management are to:

- mimic features of natural ecosystems
- increase our garden or farm's ability to withstand pests
- identify what is out of balance
- recognise common insect pests and their lifecycles
- keep pests at a level where minimal or acceptable harm occurs
- monitor the results of our intervention to see whether pest numbers are reducing and when parasite numbers are building up.

If we don't have design aims for pest management:

- we may destroy our garden or farm's best friends—the pests' predators
- we may be unaware of pest populations building up
- we may resort to using damaging chemicals.

Ecological functions of pests

Pests are species out of balance and a pest build-up is an indicator that something is wrong with your garden. Very often an infestation is a result of other factors being out of balance, such as, the soil is too dry or wet. Pests assist pollination, help create compost, recycle nutrients, disperse seed and add to biomass.

Causes of pests

- Monocultures—in single-plant crops, pests and diseases build up and spread rapidly because all their food is continuous.
- Inadequate plant nutrition and watering creates unhealthy plants that are susceptible to pests.
- Excessive feeding and watering can make plants sappy and delicious for pests.
- Plants that lack protection from hot or cold winds (which can also bring pests) are particularly susceptible.
- Plants growing in the wrong place or in the wrong season can be weaker and vulnerable to pests and diseases.
- Introduced insect species.
- Destruction of predators, such as when non-target pesticides are used.
- Catastrophe: fire, destructive clearing, flood.

Problems with chemical pesticides

Conventional pest management is usually based on prescription and chemicals. All commercial chemical preparations are non-selective, which means they can act on beneficial as well as pest species. They have several drawbacks:

- There is no one pesticide that will kill only one pest. All pesticides kill more than one animal—often the good ones as well.
- Pesticides can poison chickens, children, eggs, fish, cats, dogs and the plants we eat and/or cause cancer.
- It can be difficult to know which is the right pesticide and how much to use.
- Pesticides don't make plants healthy.
- They are expensive.

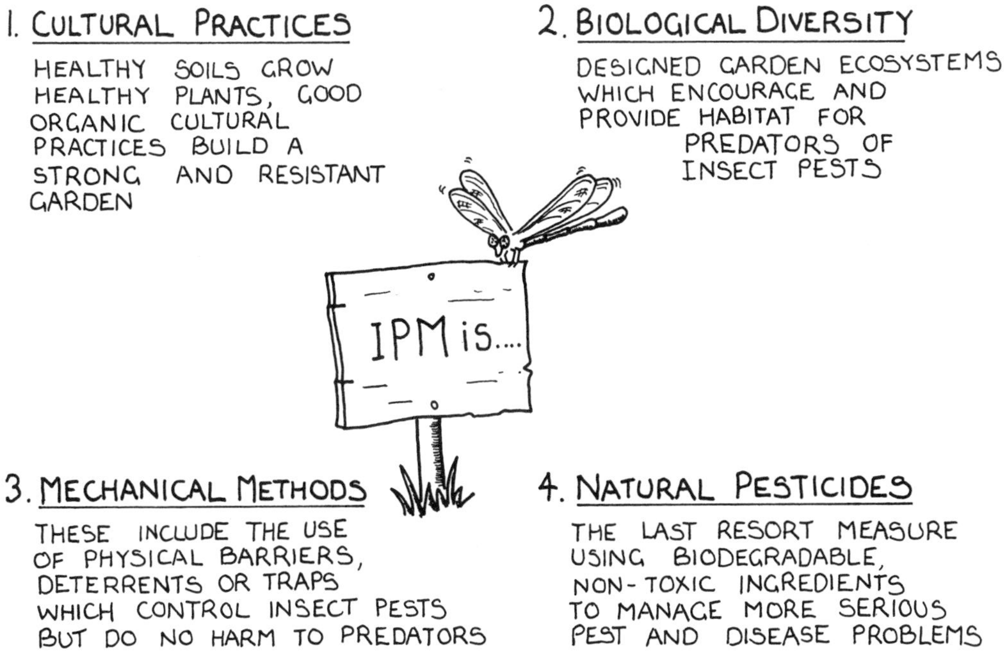

1. CULTURAL PRACTICES

HEALTHY SOILS GROW HEALTHY PLANTS, GOOD ORGANIC CULTURAL PRACTICES BUILD A STRONG AND RESISTANT GARDEN

2. BIOLOGICAL DIVERSITY

DESIGNED GARDEN ECOSYSTEMS WHICH ENCOURAGE AND PROVIDE HABITAT FOR PREDATORS OF INSECT PESTS

IPM is....

3. MECHANICAL METHODS

THESE INCLUDE THE USE OF PHYSICAL BARRIERS, DETERRENTS OR TRAPS WHICH CONTROL INSECT PESTS BUT DO NO HARM TO PREDATORS

4. NATURAL PESTICIDES

THE LAST RESORT MEASURE USING BIODEGRADABLE, NON-TOXIC INGREDIENTS TO MANAGE MORE SERIOUS PEST AND DISEASE PROBLEMS

Figure 21.1 The four strategies of integrated pest management.

Integrated pest/disease management

Pest and disease management is approached in permaculture through the integration of four main strategies which focus on working with nature (see Figure 21.1).

The four strategies are:

1. cultural practices that mimic nature
2. biological diversity to mimic nature
3. mechanical methods
4. natural pesticides.

Cultural practices that mimic nature

The best pest-management strategy is a resilient, balanced garden with:

- water in several ponds for frogs, fish and bees
- food for predators: living fences of fruit and flowering species
- homes for lizards and snakes
- poultry in orchards
- mixed herbs and vegetables
- plant and soil filters and barriers for preventing pest build-up and migration.

Cultural controls minimise pest outbreaks, especially in commercial crops, and also improve the health of your garden or farm. Well-designed ecosystems will have a variety of natural features that create ecological balance. On this land you will see the following strategies being used.

Strategies and how they work

Cover crops

Cover crops are planted sacrificially to remove or deter pests before sowing a susceptible crop. For example, mustard and rape remove wireworm in soil.

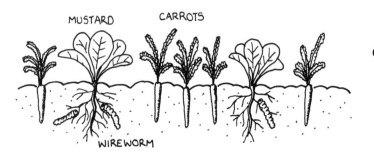

MUSTARD   CARROTS

WIREWORM

Sanitation

- Remove and destroy all diseased and fruitfly affected fruit, and leaves with fungal diseases such as black spot. Seal them in strong black plastic bags and leave them in the sun to cook. After this they can be added to the compost.
- Control weeds by hand-pulling or shallow cultivation to avoid disturbing vegetable roots. Damaged roots are an entry point for disease.

Variety selection

Different varieties of plants may display different characteristics that deter caterpillars, etc. A variety with vigorous growth may be able to outgrow damage caused by a pest. Select strong locally grown species. Check seed catalogues for heritage varieties and recommendations. Choose grafted trees grown on disease-resistant rootstock.

Crop rotation and timing

- Rotate crops to stop any build-up of soil pests and diseases. Generally, don't follow up by planting members of the same family in the same bed—don't plant tomatoes after potatoes as both are in the Solanaceae family.

- Grow early or late varieties to avoid plants maturing in the peak season when pests are often a problem.
- Winter ploughing to overturn clods can destroy over-wintering larvae. Cultivation exposes them to predator birds.

TOMATOES BEANS SWEET CORN

SUMMER

BROCCOLI LETTUCE CARROTS

AUTUMN – WINTER

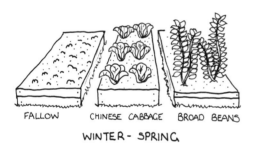

FALLOW CHINESE CABBAGE BROAD BEANS

WINTER – SPRING

Inter-planting/mixed cropping

Inter-planting varieties of plants can reduce the incidence of pests by masking the plants' colour, leaf shape and smell. Companion planting takes this idea further by planting together varieties that benefit each other. Plant aromatic herbs and companion plants with vegetables, such as tansy, pennyroyal, rue, mint, wormwood, rosemary, sage, lavender, basil, peppermint, southernwood and bay.

Rampancy and shading (plant traits)

Use pumpkins to climb over potato vine, or chokos (chayote) to climb over lantana. Dense shrubs will control weeds.

Well-managed soils

Soils high in organic content are like a factory for pest and disease control, full of beneficial micro-organisms, bacteria, fungi and even natural antibiotics which keep pathogenic organisms in check, resulting in strong healthy plants.

Regulated and appropriate watering systems

Know the water-holding capacity of your soil and the water needs of your plants. Get your watering right to avoid stressing plants. Mornings and evenings are the best time to water and remember to keep soils mulched.

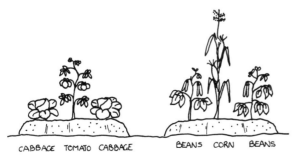

CABBAGE TOMATO CABBAGE BEANS CORN BEANS

A variety of plants

Specifically invite predators and parasites by increasing their habitat. Plant a variety of plants including groundcovers, fruit trees, vines and perennials. These attract and shelter a range of pest predators.

Biological diversity to mimic nature

This approach goes to the heart of permaculture. The primary design aim is to create gardens and farms with a variety of natural features that encourage and provide habitat for predators.

Biological control requires three elements:
- water
- food for predators
- homes for lizards, frogs, snakes, wasps and others.

You can also use domestic stock to control pests, such as putting poultry into orchards to forage for insects.

Your garden as an insect zoo

When your garden is busy and noisy with insects working in it and bees, wasps and drones all singing, then pest management is happening. Bees are possibly the most useful insect in a garden and can be killed by any and every broad-spectrum spray. Remember, your garden should be an insect zoo: the more insects, the better the pest control.

Predators in your garden will work for you and carry out the valuable work of pollination and fertilisation of fruit and vegetables. They live in the windbreaks and living fences in gardens and orchards.

Parasitic predators of pests

Insect predators often lay their eggs into a living caterpillar and when the larvae hatch they eat out the insect's insides. For example, wasps eat insect eggs and live insects. They capture insects to feed to their young. Flies attack and eat eggs of butterflies and moths. They also feed on mealy bugs of cassava.

Larger predators and what they do

Insectivorous birds are a huge help in the garden, eating grubs, lerp, aphids, weevils, bees, moths, larvae and flies. The most common small birds are blue wrens, thornbills, pardalotes and silvereyes. Because they are small they need to feel safe. Larger insectivorous birds that live in the canopy or woodlands include magpies, wattlebirds, cuckoo shrikes, friarbirds, kingfishers, ibis and willy-wagtails. They help to control pests such as Christmas beetle, psyllids, grasshoppers and locusts. Insect-eating birds eat a lot of pests. Some feed 500 insects a day to their babies. A pair of insect-eating birds can eat 5000 insects in one day.

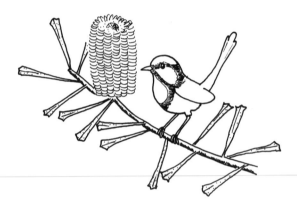

Design strategies to entice them:

Offer insectivorous birds sacrificial fruiting species (sour, non-grafted fruit trees) and several types of water (birdbath, pond or spray) because some like a plunge bath, others a dip, and some just a swoop through. For the small birds, provide safety from cats and larger canopy-dwelling birds by planting thorny, dense, small-leaved shrubs for protection; e.g. *Pyracantha* and *Japonica*, *Grevillea* and *Hakea*.

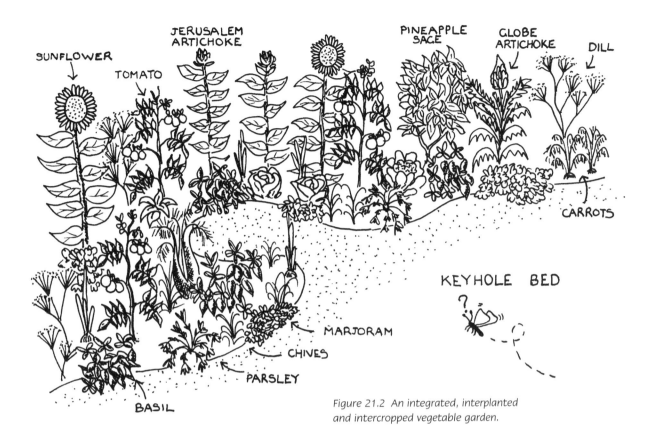

SUNFLOWER

TOMATO

JERUSALEM ARTICHOKE

PINEAPPLE SAGE

GLOBE ARTICHOKE

DILL

CARROTS

KEYHOLE BED

MARJORAM

CHIVES

PARSLEY

BASIL

Figure 21.2 An integrated, interplanted and intercropped vegetable garden.

Domestic animals are big helpers

- Birds on the ground such as hens, quail and ducks can effectively break pest cycles by eating fallen fruit. Quail need to be netted. Bantams, ducks and chickens will feed on snails, some slugs, cutworm and other insects. Guineafowl feed on grasshoppers and insects.
- Build the chicken run around the garden or beside Zone I.

- Let chickens into your orchards (Zone II) and under the house to eat white ants. It's best to keep them out of your vegetable garden because they destroy it with their feet and eat some plants; however, they can be left in a vegetable garden in movable domes.
- Fish (guppies and goldfish), toads and frogs eat caterpillars, insects, cherry/pear slug, and mosquito larvae. One frog can eat 10,000 insects in three months. Snakes and spiders eat insects. These predators need cool, moist refuges such as ponds, pipes, rotten timber logs and wetlands.

207

Smaller predators and what they do

- *Lacewings:* Lacewings are one of the most effective insect predators. As larvae they feed on aphids, psyllids, mealy bugs and moth eggs. Adults are small with gossamer wings and feed on nectar, pollen, aphids and mealy bugs. Their eggs are easily recognised sitting on separate stalks.
- *All spiders:* Spiders do a good job of cleaning up pests.
- *Ladybirds:* Both adults and larvae feed on aphids, mealy bugs, scale and other small insects. They eat 40–50 insects every day. Their babies eat more. A few ladybirds are actually pests—the 28-spot is a plant eater.
- *All wasps:* Wasps are insect predators or parasitoids. Parasitoids are more effective than predators. Parasitoid changes are often recognisable; for example, whitefly eggs go from cream to black when parasitised.
- *Robberflies and hoverflies:* Hoverfly larvae are voracious and feed on aphids, mealy bugs and mites.
- *Dragonfly larvae and damselfly:* These feed on flies and mosquitoes.
- *Praying mantis:* Praying mantis are excellent predators of many insects.
- *Ground beetles and tiger beetles:* These beetles feed on slugs and insect eggs, and their larvae feed on all other insects.

Plant allies of predators

Find a place among your vegetables and crops for these useful plants that encourage predators of known pests, repel pests or hide pest-susceptible plants. Sow them in your borders and hedges and inter-plant them among pest-susceptible vegetables and crops.

- *Family Asteraceae:* cosmos, chrysanthemum, aster, chamomile, artemesia and marigold.
- *Family Umbelliferaceae:* yarrow, anise, dill, angelica, fennel, parsley, Queen Anne's lace.
- *Other useful plants:* neem tree, garlic, dill, ginger.

Mechanical methods and how they work

Mechanical methods do minimum harm while reducing pest numbers and/or interfering with their activities. Reducing pest numbers by deterring them or manually removing them by hand does the least harm to a garden ecosystem and its resident beneficial insects and predators.

Barriers

Place fresh sawdust, sharp sand, soot, cinders or ash around special plants or beds as  barriers to pests such as snails and slugs. These barriers are abrasive and/or dehydrating. Use tar compounds on pruning cuts to prevent disease entry.

Bands

Bands made from cardboard or old cloth wrapped around tree trunks help deter crawling insects and those pests that over-winter in the soil. Grease, resins or horticultural glue can be used inside the bands to trap them. A band of lime around seedlings and trees can deter caterpillars.

Collars

Collars to deter cutworm can be made from cardboard or beer cans with both ends cut out and pushed 5 centimetres into the soil over the seedling.

Traps

Upturned citrus shells, sticky yellow boards and half bottles of stale beer are all good traps. Colours have strong effects on some insects: white attracts thrips; yellow is a good attractant for traps. Lights over water at night attract many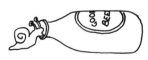

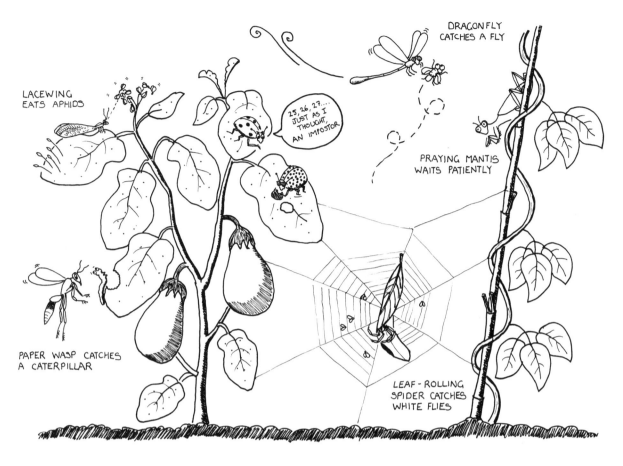

Figure 21.3 Predators quietly at work in the garden.

insects, which fall in the water and become food for fish and frogs.

Lures and baits

- Milk, like beer, attracts slugs and snails. They drink and drown, then can be collected and fed to chickens.
- Yeasts, sugars and proteins are baits for fruit fly. Put vinegar and sugar solution in traps made from plastic bottles with the tops cut off and then inverted. Ten to a tree will kill fruit fly. For this to be successful the neighbours may have to do it too.
- Pheromones are secreted by insects and can be used to manipulate the pest's

communication systems. Use synthetic pheromones in commercial traps such as dakpots or fly traps.

- Attract pests to lights or 'decoy' plants.

Exclusion

- Mosquito nets, netting fruit trees and bagging fruit all keep pests out. Make sure the flowers have been pollinated and the fruit set before putting covers on. Cheap green mosquito nets are excellent at protecting fruit from raider birds and fruit fly.
- Paper bags on rockmelons, tomatoes, eggplants and capsicum keep grubs out.

Biological

Releasing sterile males to reduce fertility has been recorded as 90 per cent effective.

Handpicking

- Handpick grubs or eggs daily—a very effective and pleasurable excuse to be outside. Catch slugs and snails at night with a torch and feed them to chickens, ducks or pigs in the morning. Pull off leaves with grubs or eggs on them and stamp on them.
- Vacuum cleaners can be effective. Hose the pest off the leaves if it cannot fly.

Desiccants

Diatomaceous earth, while non-chemical, is also non-renewable. Crushed and powdered limestone is the newest desiccant. Boric acid is a good desiccant. Borax sodium is a repellent.

Natural pesticides

Organic sprays work on insect anatomy and are usually a last-resort measure because some of them will kill beneficial insects as well as the target species. You must know what your target species is. More and more commercial preparations that reach organic standards are becoming available. To find out about the range of these you can check reliable organic mail order catalogues. Some recipes are made at home from ingredients readily at hand and use plant products. There are many recipes. Often the juice is squeezed from the leaves, mixed and then sprayed on the identified pest or disease.

TABLE 21.1: NATURAL PESTICIDES

Recipe	Target
Garlic + soap + oil	Fungi
Cassia + water	Fungi and virus
Amaranth + water	Bacteria and fungi
Leucaenia + water	Bacteria and fungi
Red onion	Fungi
Black pepper	Worms and moths
Hot pepper	Rice moth

Knowing a little about insect anatomy will help you to understand why non-chemical treatments such as 'flour and water mixed spray' work. It also helps you to think of alternative ways to manage the pests. For example, insects breathe through spiracles down their abdomens; if these are covered, the insect will be asphyxiated.

Sometimes dusts are made for soft-bodied animals. These are usually flour and salt or something that will burn the pest. Table 21.1 was compiled by Cambodian farmers. Some of the recipes have been handed down for generations.

A few more natural pesticides are examined in more detail in Table 21.2. Note that you should never use nicotine sprays as they are too toxic altogether.

Common pests

Only a very few insects in each group damage food plants so you can easily learn which ones are pests.

TABLE 21.2: PESTICIDES

Treatment	How it works
Derris dust	A general insecticide made from derris plant root; used strategically on beetles, sawfly, caterpillars, thrips and aphids.
Pyrethrum	Derived from the flowers of *Chrysanthemum cinerariaefolium* or *C. roseum*, pyrethrum is a powerful knockdown poison which breaks down in about 12 hours. It's effective against leaf-hoppers, thrips, whitefly, aphids, lice and fleas. It also kills lizards and bees.
Garlic	Crushed garlic applied as a paste is a general insecticide, useful against small and soft-bodied pests—it won't harm frogs or praying mantis.
Dipel R	Dipel R is a biological control—*Bacillus thurungenensis*—and is a specific insecticide effective only against caterpillars, the larvae of mosquitos, flies and beetles. It is useful if used strategically. Do not overuse because insects will develop resistance. Dipel R is now suspected of causing Parkinson's disease.
Bug juices	Collect some of the pest insects you want to repel and put in a blender or mash with a mortar and pestle. Leave for 1–2 days to encourage pathogens, then mix with water and use as a spray. This acts by: • releasing repellents which tell other insects to leave • attracting predators, or • spreading an insect disease. Never use for insects such as mosquitos or flies, which can carry human diseases.
Virus spray	Place five caterpillars of the pest species in a bucket of water, leave for 1–2 days and then spray on the diseased plant. Naturally occurring diseases are spread to other caterpillars.
Weed brews	Fill a 200-litre drum with 6 kilograms of cow manure and add to it six or so species of the dominant weeds in the garden. Let it brew for a couple of weeks then spray it on infested or damaged plants to act as a repellent.
Neem spray	This tropical tree is a bitter repellent for grasshoppers and locusts. It can also upset insect hormonal balance. It's available commercially.
Fungicides	Dilute urine at 1:20–30 parts in water and spray on crops or plants showing fungal diseases. Milk, powdered or fresh, is a good fungicide at 1:9 in water. Seaweed tea soaked for 2 weeks in water then washed on is another effective fungicide. 1 cup of casuarina leaves in 1 cup water, boiled for 20 minutes and then diluted 1:20, is a useful ground spray for fungal infections.

Caterpillars

These come in many sizes and colours, and some have hair. They all have soft skin and chew leaves and stems. Caterpillars hatch from eggs usually placed on the underside of leaves. Their lifecycle is:

- *Egg:* The caterpillar emerges from the egg and eats the leaves and stem of plant.
- *Larvae:* The caterpillar sleeps and is wrapped up and does not eat or drink. No damage to the plant.
- *Butterfly:* The butterfly drinks a little nectar, lives only a few days and lays hundreds of eggs. It does not damage the plant.
- *Eggs:* From out of the eggs hatch hundreds and thousands of hungry little caterpillars.

Bugs

These are sucking insects with a hard skin. Some are quite large while others can barely be seen. Their manure is liked by ants and it will also grow a black fungus called sooty mould. Their lifecycle is:

- *First-stage nymph:* The nymphs hatch from eggs. They are very small with soft skin and eat a lot. They do not lay eggs. Their skin splits and they become ...
- *Second-stage nymph:* These nymphs are bigger and larger with harder skin. They eat a lot more, but do not lay eggs. Their skin splits along the back and they become ...
- *Third-stage nymph:* These are large and strong with hard skin. They eat a lot and lay

eggs on top or under the leaf.
- *Eggs:* From these, the first-stage nymphs hatch and the cycle begins again.

Aphids

These are very small insects that have a sucking mouth. They attack leaves and stems and sometimes the flower buds. They live in groups and can grow wings and fly to other plants. When many aphids attack a plant its leaves look pale and weak.

Beetles

These are chewing insects that eat flowers, stems, leaves and even roots. There are many types of beetles and each type eats a different plant or root or seed or fruit. The beetle lays eggs which hatch into little beetles.

Borers

Borers are insects that look like caterpillars but they grow and eat and live inside the fruit or leaves or roots of a plant. As butterflies they lay eggs inside the plant.

Flies

While flies themselves do not damage plants, their eggs hatch into larvae or small worms which live inside fruit or vegetables and make them rotten.

Grasshoppers

These chew the leaves of many plants such as beans, cabbage, grass and corn, and grow very quickly. They are usually dry-season pests. Many together can eat a whole garden.

Nematodes

Very thin worms that are almost impossible to see, nematodes live in the soil and stems of plants and suck the sap. The plant gets sick because it cannot get water into its roots and up to the leaves.

Scale insects

These are very small and have a hard shell-like back. They stick themselves to stems, the undersides of leaves and flower buds to suck out the juice.

Slugs and snails

These have 300 tiny, tiny teeth which they use to eat holes in plants. They come out at night, or in damp, rainy conditions.

Solving a specific pest problem

When you have a pest problem in your garden or farm, use your observational skills to correlate the prevailing conditions with the outbreak. Over time you will build up your knowledge of critical times and pest incidence, so that you will be well prepared and the damage will be minimal. Follow these steps:
- stand, analyse, correlate and register
- identify the pest and the stage in its lifecycle
- examine the problem closely
- decide what to do.

Stand, analyse, correlate and register

Take a close look at your garden and the individual plants. Note:
- the degree of damage
- the soil conditions: whether dry, bare or wet
- the watering regimen: regular, intermittent, too much or too little
- the presence of other animals of any type
- the weather conditions—heatwave, strong winds, rainy, cold nights, frost, day length
- whether the plant variety is susceptible
- any other factor.

Identify the pest and the stage in its lifecycle

If you already know the insect then identify the stage in its lifecycle when it is most vulnerable. Carry a hand lens or magnifying glass with you in the garden to check details and then refer to books and references. There are two main types of lifecycles (see Figure 21.4):
- *Nymph lifecycle* (the gradual lifecycle): Eggs are laid which hatch out as a nymph (such as bronze orange bug) and these go through many other stages, each larger and stronger than the one before. Eggs are laid in clusters. The last stages are harder to eradicate so intervene early in the pest's lifecycle.

● *Metamorphic lifecycle* (the complete lifecycle): Eggs are laid that hatch out into a grub, maggot, larvae or caterpillar. These all then pupate into a butterfly, moth, fly, beetle or mosquito. The moth can lay hundreds of eggs per day, so try to control the insect's lifecycle before this stage.

Next, identify the pest's feeding characteristics. Insects feed in different ways and have different mouthparts. Some insert a long tube called a proboscis into tissue and suck out the contents, others chew, and some live and feed inside the fruit or leaf. Accordingly, different control methods are directed at the different ways of feeding.

● *Does it chew?* Caterpillars, snails and slugs—use stomach poisons, predator insects, baits, sprays and coarse substances or handpick.

● *Does it suck?* Aphids, thrips, scale—use desiccants (salt or flour), asphyxiation (soap, flour or salt) and predator insects, or hose or vacuum off.

Finally, determine its body type. Does it have a soft skin or hard exoskeleton? Use vinegar or salt and water or soapy water on soft-skinned pests. Feed pests with hard exoskeletons to ducks or chickens.

Examine the problem closely

● How much damage is occurring? Each pest damages a type of plant in a distinctive way, such as the leaf, stem, fruit or roots.
● Does it matter? Is it only bits of leaves and not what you want to harvest from the plant?
● Is the pest still there? Has it moved on, does it only come out at night?
● What is the pest? Get a good insect identification book.
● How big a population of them is there?
● Have the predators arrived?

Decide what to do

The final step is to choose one of the following options. They are listed in order of priority:

1. Do nothing—you can live with the damage, but do monitor the plant daily.
2. Take action. Remember to always choose the solution that will cause the least damage to the surrounding ecosystem.
3. Improve the habitat—feed (a drink of liquid manure can help the plant within 24 hours), water, mulch, remove competing weeds, shade the plant. For example, red spider mites increase under conditions of heat and dryness. A light hosing will reduce the infestation.

Figure 21.4 Insect lifecycles.

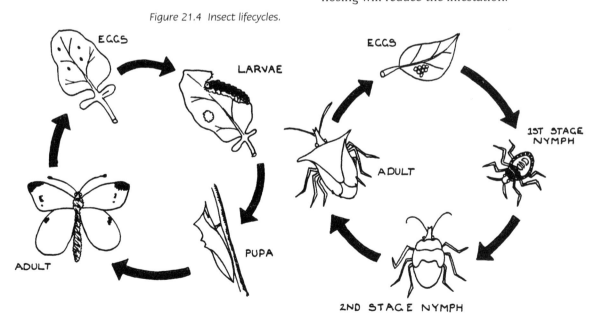

A. METAMORPHIC LIFE CYCLE   B. NYMPH LIFE CYCLE

Figure 21.5 Entice birds into your garden by providing a birdbath and keeping it topped up with water.

Common diseases

Plant diseases are caused by micro-organisms such as a virus, fungi and bacteria. You cannot see the micro-organisms but you can see the damage. Diseases are often inside the root, stem, leaf or skin of a seed or fruit. Some diseases are caused by chemical imbalances in the soils.

Viruses

These are microscopic in size and are present in fruit, leaves and stems. They live in the plant cells. Leaves of affected plants have black, white or yellow marks and the whole plant can die very quickly. Some viruses can distort the leaves, buds or petals.

Bacteria

Although larger than viruses, bacteria are still too small to be seen with the naked eye. They live in

TABLE 21.3: TREATING DISEASES

Symptom or cause	What to do
Plant colour is yellow, patchy, etc.	Feed with compost or liquid manure.
No new leaves	Water well and give compost or liquid manure.
Fed too much chemical fertiliser	Water heavily 2–3 times a day for three days.
Over-watered	Drain the water away by a channel; stop watering; add dry straw mulch; add compost; plant more plants to dry up the soil.
Too much wind and dust	Protect by palm leaf or other dry leaf; plant living fence and windbreak; wash plant well.
Monoculture	Integrate planting between plants and beds.
Plant growing in wrong place	Move it; next time find a better site.
Wrong variety of plant	Grow traditional varieties; select resistant varieties; practise integrated planting.

the sap between the cells. The entire plant or parts of it can look very sick, smell bad or go rotten.

Fungi

These are small types of mould. They have many thin white threads and some can look like white or black powder. Sometimes they have fruit, which look like small mushrooms. The plant often cannot feed or breathe well.

Effects of disease

- Plant looks sick—leaves are droopy or wilting.
- Fruit rots, is watery, smells bad.
- Leaves, seed, fruit and/or flowers drop off.
- Leaves, seeds, fruit and/or flowers have brown patches.

Treating diseases

Look carefully at the plant and its environment to identify the problem. Always choose the control method that will involve the least damage to soils, water and life.

 Try these:

1. Get a hand lens or magnifying glass and go out every day and look closely at the insects in your garden. See what they eat and what eats them.
2. Find some damage, observe it closely and describe what you think may have caused it.
3. Describe three insects in your garden, their lifecycles and decide whether they are pests or predators.
4. Describe a predator and its pest and find out what type of habitat it requires to continue to live in or be attracted to your garden.
5. Investigate one parasite and its host.
6. Which method of control do you prefer to use and how have you designed your garden to make it work?

215

CHAPTER 22

Living with wildlife

Wildlife is that wonderful range of indigenous life that lived on our land before we did and which we hope will be there long after we are gone. As natural environments are degraded so, too, are natural populations of all species, from earthworms to eagles, which provokes further environmental decline.

Wildlife is an integral part of the ecosystem we live in and we do not know which species are critical in preventing a cascading collapse. In addition, as wildlife is often in conflict with humans, it is in our own interest to provide sanctuaries for it.

Our ethical task is to:

- avoid harm to all wildlife and protect and increase populations which are threatened, rare and endangered
- practise the intergenerational equity principle, which states that we should leave all creatures and plants for future generations and it is not our role to take wildlife to extinction.

Our design aims for wildlife are to:

- encourage local wildlife for the benefits these species bring to the ecosystem
- design habitats that are animal friendly
- employ life-affirming strategies, such as not using any pesticides on our land
- create wildlife sanctuaries
- apply deterrence strategies to keep garden raider animals such as possums in check.

If we don't have design aims for wildlife:

- animals can become locally extinct
- crops can be destroyed, wasted and infested
- wildlife populations build up and create imbalances
- new immigrant wildlife can move in and compete for food and habitat
- feral and pest species can invade.

Ecological functions of wildlife

- Pollination of fruits and flowers by wasps, birds, bees and small mammals.
- Predation of insect pests.
- Seed dispersal and germination through transport on paws, wings, fur and eating.
- Grazing and browsing manages pastures and weeds and encourages bushiness.
- Cleaning of land, rubbish, and decayed materials.
- Tractoring to clean land of unwanted plant cover.
- Diversity to build strong ecosystems and to support each other.
- Connection in hubs, nodes and keystone

Figure 22.1 Roadside vegetation provides important wildlife corridors and habitat.

species and with us. Keystone species are those upon which a whole inverted pyramid of life depends. For example, bees are keystone species. They pollinate fruit trees, herbs, and a multitude of other plants, and we and other wildlife all depend on their fruits. Some plants are also keystone species. In Australia, *Banksia ericifolia* flowers in winter when there is little other food, and provides for birds, small mammals, insects. Without these flowers, many animals would become extinct. We often don't recognise keystone species.

Strategies and techniques for encouraging wildlife

You need to use the principle of friendly encouragement, which requires the creation of natural, wildlife-friendly habitat. Ultimately, you can't make wildlife come to your land and stay unless you provide homes and habitat. Set up the conditions under which the habitat can evolve and you will find the wildlife will move in. It is a joy to see.

Figure 22.1 shows the effects of maintaining roads as wildlife corridors. Sometimes these corridors are the very last little bit of habitat left to endemic animals and plants.

Provide wildlife corridors

Wildlife corridors are vitally important and highly significant because they allow animals and plants to move and find new habitats when theirs are under stress from building, fires, clearing, spraying, etc. Often these are the only habitats some species have left.

Plant along roadsides, windbreaks, harvest forests, Zone V, and projects that connect small reserves with national parks; plant corridors on farms along rivers, creeks and farm roadways.

Provide and maintain water

Give as many types of water as possible: from deep to shallow, from still to running, exposed and with overhanging vegetation. Animals use water for drinking, bathing, food supplies, hiding places and cooling.

All animals must have water and some need it several times a day. Many become dependent on the water source and will die if it dries up. Many live permanently very close to it. Only 26 per cent of Australia's birds are nomadic so the other 74 per cent must have a permanent water supply. Many waterbirds are migratory. They require a hectare of surface water to be attracted to an area, hence the importance of wetlands.

Other animals hunt around permanent water so they must have a variety of habitats surrounding it. Put dead logs in water, rocks on the side as shelving, sand, and reeds for breeding birds. Design dams that back onto forest or create forest around dams for protection.

Plant suitable vegetation

Use spiny and prickly plants to protect small birds and other animals from predators. Leave old trees in place as some wildlife (such as sugar gliders and kookaburras) can only nest in old, dead or half-dead trees. Some animals have permanent homes in them.

Ensure a dense under-canopy for small birds to live in because hawks, eagles, magpies and currawongs are tall-tree/cliff nesters that prey on small birds. Plant special insect-eaten trees that insect-eating birds will be attracted to, such as *Acacia melanoxylon*. Plant specific eucalypt trees for animals such as koalas and flying foxes that have few dietary choices.

Enticing wildlife to work for you

Wildlife can be a form of effective pest control. To encourage pest-eating birds and other wildlife to your garden or farm, try the following:

Figure 22.2 Hollow logs placed in trees create shelter for possums.

- First and always, plant habitat.
- Offer small amounts of grated cheese and fat in winter to small birds such as grey thrush, which will in turn eat caterpillars and slugs. They require shallow, hollow logs or stumps to nest in.
- Attract insects using yellow colours in the garden; robins, honeyeaters and grey thrush will be attracted to eat the insects.
- Pineapple sage and grevilleas growing in orchards will attract honeyeaters.
- Spray Biodynamic 500 outside cropping fences and kangaroos and wallabies will come to graze.
- Wallabies can be encouraged with pollard and ground oats, as can wombats.
- Echidnas like milk and a deep litter mulch in hedges and bushland.
- With new plantings there are often no nesting places for animals, so encourage wild ducks by placing 5-gallon drums open at one end near water.

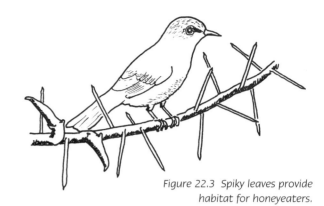

Figure 22.3 Spiky leaves provide habitat for honeyeaters.

- Piles of stones and stone walls bring a range of lizards to eat snails and slugs.
- Tadpoles like small and larger ponds of still water.
- Lyrebirds and brush turkeys need deep litter and mulch to scratch around in.

Feeding wildlife

This is a contentious issue and in countries like Canada and Australia conservationists do not want you to provide food for wildlife because the animals:

- build up a dependency on being fed and so don't seek their own food
- tend to prefer unnatural foods which are bad for them (the wrong foods can cause animals to sicken or die if they overeat them)
- become tame, which makes them vulnerable to ruthless people.

However, it is sometimes appropriate to offer food to wildlife when it is endangered by:

- land clearing: when a large number of exotic trees were removed from my place to make way for a diverse planting of indigenous

Figure 22.4 Rocks and overhangs make your garden lizard friendly.

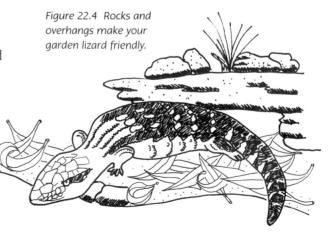

TABLE 22.1: KEEPING RAIDER ANIMALS IN CHECK

Strategies	Techniques
Decoys—entice the animal to go elsewhere	Sow small patches of turnips or buckwheat as decoys around wanted crop.
Deterrents	Strong smells repel some animals: • sprinkle coffee grounds • inter-plant aromatic herbs, such as mint. • use old tyres because the smell of rubber can act as a deterrent. Plant crops that are unpalatable for birds. Try scaring raider animals away by hanging silhouettes of hawks, flags, old CDs, singing wires or plastic snakes in fruit trees to deter birds; or use an electrified human dummy to deter tigers. To be effective these are put up as crop ripens and removed immediately afterwards.
Mechanical barriers	Electric fences for large mammals, angled fences for jumping and climbing animals, gravel paths for snakes.
Other	Make a productive crop or enterprise from the raiding animal; for example, harvest rabbits, kangaroos and deer.

plants for habitat, some animals lost food supply and so I fed them until they adjusted to the new situation—about twice a week for six months, tapering off as they found new supplies and the plants grew

- extreme weather conditions such as flood and extreme cold
- the loss of habitat after bushfires
- any loss of habitat until it is re-established.

If you must feed them, then take the trouble to prepare the following dry food mix:

3 cups baby cereal

1 cup rice flour

1 cup glucose powder

1 cup egg-and-biscuit mixture (from pet shops)

1 teaspoon multivitamin mixture (from pet shops) (optional)

half an orange or apple or a piece of other fruit (optional)

Strategies for keeping raider animals in check

The principle of deterrence, which is to deter animals from crops rather than kill them (also known as the 'go somewhere else' principle), is now widely accepted and research on deterrence has yielded some good results. You can introduce

deterrent-effective strategies and techniques into your garden. Many strategies for keeping raider animals in check are covered in Chapter 21 on pest management and also apply to wildlife. Some people tell raider animals very firmly what is theirs and to stay away from the rest (see Table 22.1).

Try these:

Get a separate workbook or use a special section in your observation journal for wildlife. You will find that your observation and understanding of your ecosystems will improve enormously.

1. List any animals you would like and/or need in your garden or on your farm and describe how you will encourage them.
2. Design special animal-attracting features on your land. Draw the design in your book. What animals will be attracted by these designs?
3. Make an inventory of animals seen on your land. Do this again after a year or so when your garden is well established.
4. Check how successful your design is each year at a regular recording time.

CHAPTER 23

Weeds: guardians of the soil

Weeds have been badly maligned. They have earned their bad reputation because they compete with agricultural crops, and some farmers and gardeners restrict the plants they consider economic or aesthetic to a very narrow spectrum. Paradoxically, however, there are some cultures that do not even have a word for weeds in their vocabulary because for them every plant has a use.

Recently ecologists and other scientists have reviewed the idea of weeds as nuisances and have examined the role and function of weeds in natural and cultivated ecosystems. They now consider weeds to be clear environmental indicators that carry out a variety of valuable functions as well as helping in the diagnosis of land problems. A large numbers of gardeners and commercial farmers manage weeds very well without chemicals by following certain principles, strategies and techniques which are often locally appropriate and proven.

 Our ethical task is to:
- appreciate weeds for the way they function in systems and not remove them until we have something better to replace them with. Remember, soils don't like to be naked. Weeds rush to clothe them.

 Our design aims for weeds are to:
- work with weeds to get closer to your land, to observe, deduce and interact
- keep weeds at a manageable level; don't make weeding a chore, it is a pleasant way to spend some time; find ways to harvest and use weeds

- look at weed problems as a fascinating study with an intrinsic solution.

 If we don't have design aims for weeds, they:
- compete with useful plants for water, nutrients and light
- poison humans and domestic animals
- can taint agricultural produce, such as milk
- may harbour disease and pests
- interfere with transport and essential services; for example, water weeds, weeds in sewage
- are parasitic on other crops
- invade or pollute ecosystems
- can be unsightly.

Ecological value and functions of weeds

Weeds try to restore ecosystems and their presence tells us something has gone wrong. They:
- act as indicators of soil pH and nutrient status
- improve soils by changing the soil pH
- provide nutrients for impoverished soils
- mine minerals by extracting them from lower

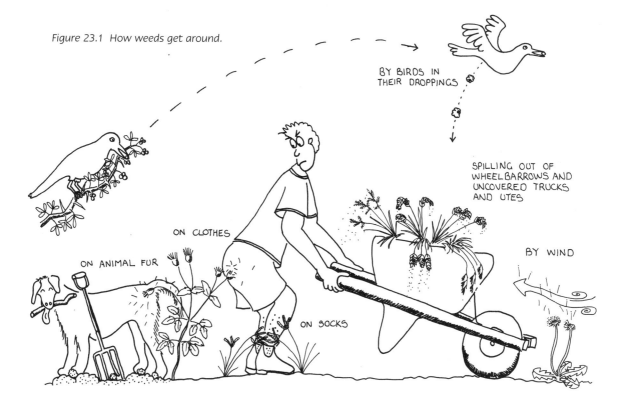

Figure 23.1 How weeds get around.

soil strata and bringing them to the surface in their leaves; these rot and the minerals are then available in organic matter to other plants

- reduce water loss by acting as living mulches
- restore succession processes preparing soils for the next stage
- loosen hard, compacted soil and absorb surplus nutrients and water
- are colonisers, being the first to arrive and live under harsh conditions
- protect damaged soils from grazing animals by toxic or thorny weeds
- cover loose soils, preventing them from being eroded or blown away
- exhibit rampancy, which enables soil or other plants to be covered very quickly and strongly
- add to landscape diversity and strength, and are often habitat for predators and wildlife.

TABLE 23.1: WEEDS

Conditions favouring weeds	Type of weeds
Overgrazing—nutrient depletion, compaction	Thorny or toxic plants gain hold to remove animals and protect soil.
Fire and clearing	Tenacious and hardy plants establish themselves.
Chemical soil changes—very diminished nutrients	Sedges and sour grass arrive which can tolerate these conditions.
Loose, bare soil from ploughing, clearing, hoeing	Crops of annual exotic plants and especially grasses attempt to cover the soil.
High nutrient availability and increased moisture from irrigation, garden run-off, farms and over-watering	Tenacious exotic tree and shrub species establish themselves.
Increased light intensity	New plants arrive to invade, compete and thrive.
Exhausted soil	Mosses and thistles emerge.

Figure 23.2 Four examples of conditions that favour weeds: overstocking, nutrient run-off, land clearing and soil disturbance.

OVERSTOCKING AND OVERGRAZING LEADS TO COMPACTED, BARE SOILS WITH LITTLE NUTRIENT

INCREASED WEED GROWTH

HIGH NUTRIENT AVAILABILITY AND INCREASED WATER RUN-OFF FROM BACKYARDS INTO URBAN BUSHLAND

WEED GROWTH IN CLEARING

INCREASED LIGHT INTENSITY

SOIL DISTURBANCE AND WEED SEED ON MACHINERY OR ON IMPORTED MATERIALS, FOR EXAMPLE IN ROAD CLEARING AND MAINTENANCE

Why we have weeds

Usually weeds appear when successful and stable ecosystems are altered so that the new conditions favour them. Weeds thrive in disturbed and poorly managed conditions. Each condition results in a different group or type of weed invasion. Table 23.1 is a guide to the types of land disturbance that encourage weed infestations and the types of weeds that thrive under the altered conditions.

As an observant permaculture designer, you will look at any of these situations and forecast weed invasion. You will also know how to avoid conditions favouring weeds.

Weeds and succession

Weeds are colonisers and grow in succession; left to themselves they will usually die out and be succeeded by climax species. For example, grasses are succeeded by herbs and then by nitrogen-fixing species. In many cases the original ecosystem will eventually return because it is most suited to the interacting combined environmental variables of wind, soil, rain, temperature, insects and so on. This depends on birds and animals bringing in seed, or having a large bank of seed in the soil.

Weed management strategies

Today the term 'weed control' is really just a euphemism for using herbicides. However, there are six cogent reasons not to use them:

- they impact on human health
- they impact on the environment
- they impact on garden and farm management
- the long-term effects are often unknown
- they create dependency, mutations, and may be withdrawn one day
- they do not assist plant and soil nutrition.

By developing an awareness of your land and its needs, you can develop effective weed-control strategies that avoid the use of herbicides completely.

Know your land and bioregion

You can spend a lot of time trying to eradicate a plant that may never become a problem. It is also possible to miss a plant that will become a problem. Observation is crucial when you work with weeds. Observe climate, temperature, light changes and weather; for example, a wetter than normal summer, or a colder than usual winter. You will find that different weed crops respond to the different conditions. When you know your microclimate you can predict which weeds will germinate.

Know your weeds and their lifecycles

Weeds have lifecycles and seasons like all other life forms. Most plants will not become big problems and are simply colonisers early in plant succession and have short life spans—usually annual or biennial. If you are able to leave them, the next crop of plants will take over and succeed them.

Learn to recognise weeds local to your area and study their characteristics. For example, *Correopsis* is a weed that requires full sun so it thrives along roadsides and is highly unlikely to invade bushland. Kikuyu grass can be weakened and managed by frost and shade. Knowing your local conditions and associated weeds will enable you to determine whether management is needed and, if so, what strategies will be most effective. Walk around your land, observe the weeds and consider:

- Why is this weed here?
- Is it carrying out valuable functions to protect or restore the land?
- What is its lifecycle?
- How fast is it spreading?
- When is it important to work on controlling it? Mark the time period on your calendar.
- What is the best strategy and do you have the tools and means to do it?

If you take the time to answer these questions you will save on maintenance and labour.

Weed impact assessment

You carry out a weed impact assessment when you are faced with a site which is derelict or neglected,

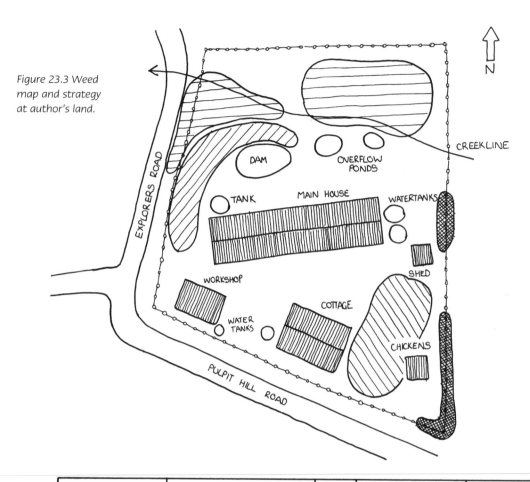

Figure 23.3 Weed map and strategy at author's land.

WEED DENSITY	WEED TYPE	HEIGHT	CAUSES	STRATEGY
90% WEEDS	• JAPANESE HONEYSUCKLE (Lonicera japonica) PERENNIAL CLIMBER	2M	GARDEN ESCAPE FAVOURING MOIST SOILS	REGULAR CUTTING-BACK, DIG OUT STEMS AND ROOTS.
50% WEEDS	• YORKSHIRE FOG (Holcus lanatus) PERENNIAL GRASS • SHEEP SORREL (Rumex acetosella) PROSTRATE PERENNIAL HERB	1M ·4M	INCREASED LIGHT AND GROUND DISTURBANCE AFTER TREE REMOVAL. DEEP ACID SOILS.	SLASH SEEDHEADS IN SUMMER AND MULCH. PLANT HEAVILY WITH NATIVE UNDER-STOREY. WOOD-DUCKS GRAZE SORREL.
35% WEEDS	• KIKUYU GRASS (Pennisetum clandistinum) PERENNIAL GRASS	·3M	LIGHT, MOISTURE AND WARMTH	MOW HARD AND OFTEN. SHADE OUT WITH DENSE PLANTING.
20% WEEDS	• BLACKBERRY (Rubis fruticosus) PERENNIAL CLUMP WITH LONG ARCHING CANES	1·5M	HIGH SOIL NUTRIENT DUE TO HORSES AND COWS. MOIST COMPACTED SOILS WITH LOW pH.	DIG CLUMPS IN AUTUMN BEFORE FRUIT RIPENS

or when natural vegetation is infested with weeds. A weed analysis will help you restore the land to good health with maximum effectiveness and minimum cost, energy and the use of other resources.

Draw two sketch plans of the infested area; one as a bird's eye view and the other as a cross-section. The author's sketch plan is shown in Figure 23.3 with an analysis below it.

- Show the height of weeds, trees, shrubs and groundcovers.
- Show the extent of infestation; for example, 30, 50 or 80 per cent.
- Estimate the probable causes of infestation: moisture, light, nutrients, etc.

Now decide how the major weeds are propagating themselves so you know how to prevent further spread. Look at the physical conditions which caused the problem and try to reverse them. There are several ways you can go about this.

The minimum disturbance strategy

This is the best strategy to use. It relies on the principles of ecological succession that were discussed in Chapter 3. If you don't remember what these are then it is important to revise them, as they are fundamental to this method.

1. Leave weeds until you have more useful, or local, plants to replace them and until you have a management strategy because bare ground is a disaster.
2. Plant the next succession plants amongst the weeds. So if you have a grass weed problem then plant herbs and nitrogen-fixing plants in amongst the grass.
3. Slash frequently, especially if you have seeding annuals. Rake the seedheads if mature and compost them or make silage. Or graze lightly with animals such as pigs, donkeys and poultry.
4. Later add, say, fruit trees and timber trees.

Remove seeds

Many plants are prolific seeders. Some plants can produce 20,000 seedlings over a season if they all grow, so a strategy that removes flowers before seeds form is very effective. Grasses can be controlled simply by preventing seeding. It is often sufficient to slash the grasses once or twice a year at flowering and before seed-set to achieve long-term control.

Use shade

Shade is a powerful way to manage weeds that thrive in sunlight and open space. Even some of the most intractable weeds, such as kikuyu, can be shaded out. You can use mulches of various types or you can mow hard and plant pioneer species really closely.

Change the conditions

Ask yourself why the weeds are protecting this land and then change their conditions.

1. Drown them out—few plants can tolerate more than seven days of flooding.
2. Dry them out by starving them of water temporarily; for example, buttercup and ranunculus can be removed by draining an area (directing water away).
3. Mulch with living or other mulches to exclude light. Use the rampancy of plants such as pumpkin to cover and exclude light. This method is very effective and prevents seeds from germinating.
4. Introduce weed diseases (see Chapter 21). There are several biological controls for weeds, such as fungus to weaken blackberry, which the Department of Agriculture or Land Management can help you with.
5. Use heat. You can burn weed patches by building very hot fires over them. Cover temporarily with black plastic in summer and allow it to cook the weeds below it. Incidentally, you can also place small amounts of intractable weeds in black plastic bags and put them on concrete or the roof and allow them to cook before placing them in the compost.
6. Change land use: turn grass into forest. Graze weed-infested forest.
7. Change soil conditions by adding organic

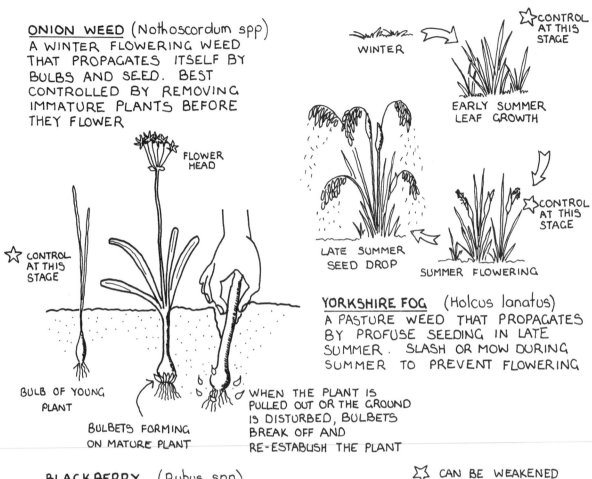

ONION WEED (Nothoscordum spp)
A WINTER FLOWERING WEED THAT PROPAGATES ITSELF BY BULBS AND SEED. BEST CONTROLLED BY REMOVING IMMATURE PLANTS BEFORE THEY FLOWER

FLOWER HEAD

CONTROL AT THIS STAGE

BULB OF YOUNG PLANT

BULBETS FORMING ON MATURE PLANT

WHEN THE PLANT IS PULLED OUT OR THE GROUND IS DISTURBED, BULBETS BREAK OFF AND RE-ESTABLISH THE PLANT

WINTER

CONTROL AT THIS STAGE

EARLY SUMMER LEAF GROWTH

CONTROL AT THIS STAGE

LATE SUMMER SEED DROP

SUMMER FLOWERING

YORKSHIRE FOG (Holcus lanatus)
A PASTURE WEED THAT PROPAGATES BY PROFUSE SEEDING IN LATE SUMMER. SLASH OR MOW DURING SUMMER TO PREVENT FLOWERING

BLACKBERRY (Rubus spp)
A WEED OF FERTILE, MOIST SOILS THAT PROPAGATES ITSELF WHEN LONG CANES TOUCH THE GROUND AND TAKE ROOT. ALSO SUCKERS FROM DISTURBED ROOTS AND BY BIRDS EATING FRUIT. KEEP LARGE CLUMPS TO A MANAGEABLE SIZE AND DIG OUT CLUMPS IN AUTUMN AND WINTER

CAN BE WEAKENED BIOLOGICALLY BY AN INTRODUCED RUST

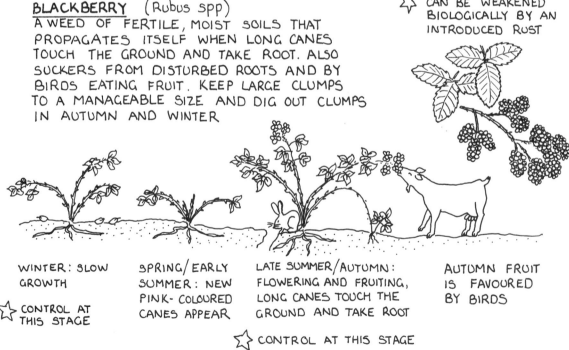

WINTER: SLOW GROWTH

CONTROL AT THIS STAGE

SPRING/EARLY SUMMER: NEW PINK-COLOURED CANES APPEAR

LATE SUMMER/AUTUMN: FLOWERING AND FRUITING, LONG CANES TOUCH THE GROUND AND TAKE ROOT

CONTROL AT THIS STAGE

AUTUMN FRUIT IS FAVOURED BY BIRDS

Figure 23.4 An example of three different weed types and the best stages of intervention.

UNPRODUCTIVE FIELD WITH
PATTERSON'S CURSE OUT-
GROWING GOOD PASTURE

SLASH/MOW REGULARLY AND
PLANT PRODUCTIVE TREE SPECIES

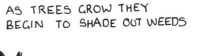

AS TREES GROW THEY
BEGIN TO SHADE OUT WEEDS

MATURE TREES PROVIDE 80-90% SHADE
COVER, ENOUGH TO KEEP WEED SPECIES
UNDER CONTROL AND PROVIDE FODDER
AND OTHER USEFUL YIELDS

Figure 23.5 Field of Patterson's curse outgrowing pasture. Remedy by supplying more organic matter and then plant browsing shrubs to shade it out using successional planting strategies.

matter, or change the pH by planting different plants, or apply one-off doses of, say, gypsum or sulphur.

8. Weaken weeds by cutting or mowing very often and severely.

Try these:

1. Carry out a weed inventory of your site. Do it at least annually but it's better done half yearly. Record the local weather conditions at the same time. Monitor the changes in weeds each year from this base data.

2. Make a table of your proposed control strategies. List all the weeds that are a problem or are environmental weeds. Your table should look something like Table 23.2.

3. Which weed troubles you most? Describe it and its habitat. What control methods will work best? Do a management plan for it each year and monitor your progress.

TABLE 23.2: CONTROL STRATEGIES

Time	Weed	Control strategy
Late spring	Annual grasses	Slash
Midsummer	Blackberry	Cut canes and lift root clumps
	Agapanthus	Cut off all seed heads
Late autumn	Honeysuckle	Pull back runners
All year	Garden invaders	Mulch with straw

4. Find a piece of badly infested land and do a study of it. Carry out a weed impact assessment. Design a strategy for repair. If you can, implement it for a year.

5. At all times record your observations about everything else (insects and birds), even things that seem irrelevant.

CHAPTER 24

Aquaculture: the water permaculture

In permaculture, aquaculture systems are regarded as water polycultures; that is, they are cultivated water ecosystems. All the aquatic organisms, including water plants, fish, crustaceans and water birds, are interdependent on each other and their environment through food chains and food webs. And, like all stable ecosystems, these cultivated aquacultures are sustainable and highly productive.

Some of the best cultivated aquaculture systems have been developed in South-East Asia. In Vietnam there are specialised aquacultures for coastal waters, along estuaries, in delta canals, in freshwater mountain lakes, and in the home fish-ponds that are an integral part of their sustainable gardens and farms. These highly complex water systems have been developed over thousands of years and are excellent models for developing similar systems in other parts of the world.

In other societies, however, most people buy fish that have been harvested from the sea or raised in commercial fish farms. There are several good reasons why you should avoid buying fish caught or raised using these methods. Firstly, an ever-increasing number of unsafe chemicals are now found in seafood and freshwater fish. Secondly, continual harvesting from lakes and oceans is causing serious environmental damage. In some areas, divers have described the seabeds as desolate moonscapes! We know that the ocean stocks are collapsing and 30 per cent of stocks are already gone. You are helping to preserve the ocean and river wildernesses when you do not buy commercial seafood and freshwater fish.

Commercial fish farms, where one species of aquatic animal (fish, mussels, crayfish, etc.) is raised in a monoculture system, are not a satisfactory alternative to harvesting from natural systems. Fish farms have all the same problems as other monoculture systems: they require high energy inputs and maintenance, and are supported by a variety of chemicals.

You will find there are many other benefits to be gained from developing sustainable aquaculture systems. You may feel it is not ethical to eat red meat because of the environmental cost of feeding most of the world's grain to beef and pigs while many people are starving, or you may be concerned about land degradation occurring through removal of forests for grazing land. Aquaculture is potentially less damaging to the environment, and therefore is a more ethical way of obtaining animal protein.

Aquaculture is also one of the most efficient methods of obtaining high-quality animal protein.

- It can be carried out on marginal land, making it productive.
- It shortens the food chain, especially when species with low trophic levels are chosen, such as yabbies, herbivorous fish and mussels.
- Fish are cold-blooded and do not use energy for body warmth, therefore they require less food per body weight.
- Fish devote more food energy to growth than land animals because their weight is supported by water.
- Fish can feed on organic wastes such as plant and animal residues.

Water from fishponds can be used for irrigation, as a nutrient-rich liquid fertiliser, and for fire-fighting. Water systems in gardens and on farms can also add to microclimate variations (see Chapter 5).

In temperate and tropical systems more aquaculture systems are needed to ameliorate environmental stress because they absorb excess nutrients, filter some toxins, and supply a wider range of cultivated products.

Our ethical task is to:

- reduce the harvest pressure on wild fish stocks by raising animal protein ethically
- add to land storage of water
- restore and increase fish stocks wherever possible.

Our design aims for aquaculture systems are to:

- site ponds appropriately in home gardens and on farms

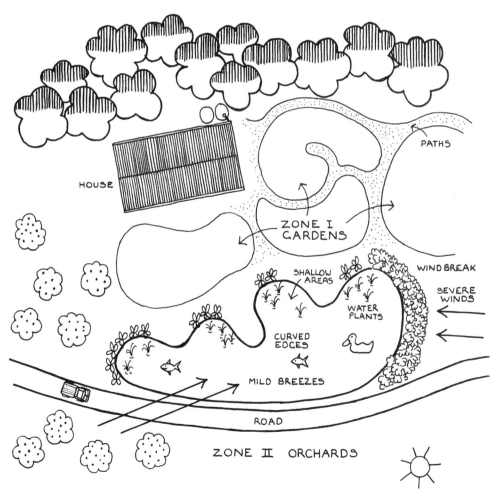

Figure 24.1 Siting of aquaculture at Rosie's farm.

Figure 24.2 Cross-section of an aquaculture dam.

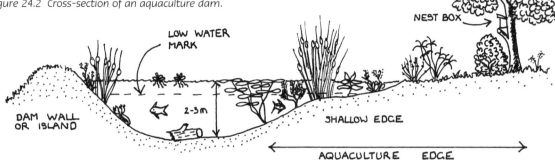

LOW WATER MARK

NEST BOX

DAM WALL OR ISLAND

2-3m

SHALLOW EDGE

AQUACULTURE EDGE

- meet criteria for setting up small and large systems
- develop strategies for managing ponds.

If we don't have design aims for aquaculture systems:

- we can create a polluting system of fish and water
- we will miss out on all the values, duties and functions of water
- we will be more vulnerable to disaster and environmental change.

Siting aquaculture systems

In a permaculture design, aquaculture ponds are placed downhill from structures and cultivated areas. The ponds will filter biological pollutants in water run-off and mop up and produce valuable products from biological excess. In effect, this filtering process closes the permaculture system.

You will also need to consider other functions of water bodies when you select your site. When placed below or close to homes, dams reflect light, cool hot summer winds, and add warmth in winter.

Dams also act as a significant barrier to fire, and the nutrient-rich water can be gravity-fed to other enterprises. All ponds should be placed where they will receive sunlight and exposure to breezes. Sunlight helps the micro-organisms to grow and breezes assist water oxygenation. If you look at Figure 24.1 you will see how Rosie's aquaculture dam is exposed to the cross-breezes and sunlight. She has positioned her dam so that the winter sun can be reflected into the house. The trees and the house at the back of the dam protect the water from excessive evaporation caused by the severe winter winds. Do not plant the edges of the pond so thickly that the sunlight and breezes are blocked.

Small water systems can be placed everywhere throughout the garden. They change the humidity and light around their immediate area, and an array of water plants does well around them. They keep cold gardens warmer and attract predators that need water regularly. Smaller ponds may become too hot during summer and should be oriented to minimise solar radiation. Shade from nearby trees will also help keep temperatures down.

Farm dams are designed to take grey water and so are placed downhill from the house. The pond

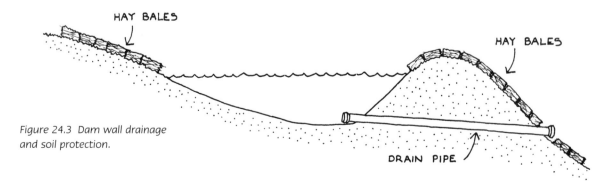

HAY BALES

HAY BALES

Figure 24.3 Dam wall drainage and soil protection.

DRAIN PIPE

filters biological pollutants out of the water run-off before it's returned to rivers or used on crops. These shallow dams moderate the climate and act as significant firebreaks. Think about evaporation from hot winds and whether there is sufficient sunlight over 60 per cent of the surface area. Sunlight helps micro-organisms grow and light oxygenates the water. Generally, water from aquaculture has an excellent pH level and is best used for tree crops. Sometimes there is too much nitrogen for vegetables.

Many smaller water systems are better than one immense one. However, the size of the water body depends on the size of the landscape, climate and available water.

Construction

Typically, farm dams are deep and narrow: the small surface area helps to reduce water loss through evaporation. However, there is no point in building aquaculture dams deeper than 2 metres because most freshwater species do not use deep water. Figure 24.2 shows a cross-section from the deep water to the shallow edge and above the water line. Many water plants grow at a specific depth and this 'shelf' adds to the planting potential of the whole system. The edge of the dam is important because it will need to support a variety of plants and offer different habitats for the animals. Aquaculture dams should have wavy and convoluted edges to increase the surface area of the dam.

The pond bottom is very important in the biology of dam life. A good bottom quickly recycles nutrients and makes them available, while on a poor bottom decay is slow. Gravel, clay and sand bottoms can be improved by adding organic matter such as stable manure, sewage sludge or sowing a green manure crop before filling the dam with water. The bottom should be free of too much silt so that when it does build up it can be removed and used as fertiliser. You should plan to manage it by cleaning it out regularly. When digging your pond, insert a pipe in the dam wall so you can drain it if predators get in or if the water becomes toxic (see Figure 24.3).

Planting

Plants growing on the edges and in the water are the backbone of aquaculture systems as they hold the soil, recycle nutrients, shelter animals, purify and cleanse the water, and can be harvested for our purposes. As soon as the dam or pond is dug, the exposed edges should immediately be planted to prevent rain washing the soil back into the dam. Hay bales can also be laid on the bare soil to prevent soil loss through erosion. The edges of small ponds can be planted with mints, including Lebanese, Vietnamese and common garden mints, and other water-loving plants such as blueberries. Clovers, lucerne, herbs and bulbs can be planted around the edges of larger dams. Some shrubs and herbs for the edges are comfrey, sweet potato, lavender, lemon grass and tea-tree.

It is also a good idea to grow fruit trees such as mulberry trees (in warmer areas) or quinces (in cooler areas) fairly close to the water—the water animals can eat the fruit that drop in the water as well as the insect pests that are attracted to the trees. Bottlebrush to attract insects, fruit trees, avocado (on the southern edge where the microclimate can be 4–5°C higher than the surrounding area) and whatever grows locally in wet areas are all good choices. Closer to the water's edge, reeds can be planted to provide sheltering places for water animals—reeds are especially useful for sheltering newly introduced fry. Seepage areas can be used for mints and trees such as poplars, willows and pecans.

When you are planting your aquaculture system, you should practise the same principles of abundance that you used in your food garden; that is, you should plant densely and introduce a wide range of desirable species. The species that are best suited to the environment will survive and eventually become naturalised. Dense plantings will also help avoid weed problems.

Choose plants for specific conditions, such as:
- running or still water
- able to grow on edges
- have edible fruits or roots
- have a high nutritional value
- have an excellent market prospect.

It is important not to introduce plants that can become weeds of waterways. In Australia, water hyacinth is classified as a noxious weed because it rapidly forms a dense mat over the water surface. In Vietnam, where the same plant is a valuable part of the aquaculture system, it is confined to small areas of the home garden ponds and is regularly harvested for mulch and pig feed. It is also used to cleanse the water.

Introducing fish

Large dams can be used to raise good-quality eating fish. You will need to wait three to six months after the dam has been built before you introduce fish to the system. This allows time for the water to settle and the plants to become established.

Different depths are important in providing a range of different habitats; large fish naturally move to the deeper water and deep water ensures a region of lower temperature during the summer months. The ideal temperature range is 18–25°C. Shallow water supports some weeds, which offer protection to small fish and are a source of large quantities of food. Shallow water plants also provide habitat for waterfowl and yield crops such as water chestnuts, taro and arrowhead.

The pH should be between 7 and 9 and will change at different times of the year. Anything under pH 6.5 won't be very productive. A pH between 7.5 and 8.0 is optimal. Smaller-growing fish are stocked first because it will take time for the system to provide enough food for them. Gradually, insects, bugs, frogs and other animals will come to live in this new housing estate and will supplement the fishes' diet.

If the system is fairly large it could take as long as two or three years before you can introduce larger fish for eating. These should be bought while they are still young so they don't immediately eat all the smaller fish.

Your first choice should be indigenous fish. In the past, inappropriate selection and management of exotic fish has caused considerable damage to natural waterways. Fish are usually specific to different

environments. There are indigenous fish adapted to salt water, brackish water, still water, coastal waters, mountain areas and inland rivers. Local fisher people in your area can tell you which species are good for eating and the conditions they require.

The size, number of fish and carrying capacity is most closely related to the surface area and not to the depth of water or total volume because fish feed at the edges more than in deep water or water far from the edge. As a very rough guide, you can stock up to 100 adult fish of 1 kilogram weight for each 1 million litres of water. However, this differs considerably for different fish in different regions and depends on the food chain. Stock prawns or varieties of local crustaceae as well as the fish and bivalves in small dams. The following are some Australian species:

- *Murray cod:* 200 fry per hectare; 120 fingerlings per hectare in cages. They like to breed under things or on floating rafts, logs and clay pipes.
- *Golden perch:* 300 fry per hectare. They are tiny fry.
- *Silver perch:* 160 fingerlings per hectare.
- *Catfish:* these are natives and dam breeders and very good eating.
- *Red fin:* this is a good small breed if it cannot escape.
- *White amure:* this is a fine eating fish. It is a plankton feeder.
- *Prawns:* these are eaten by big frogs.
- *Goldfish:* these eat mosquitoes.

Note that plankton eaters—rainbow trout and white amure—can only breed in running water, through flow forms.

Eels are very common in farm dams on the coastal side of the ranges. They eat fingerlings and so significantly reduce the chances of establishing fish in a dam. However, they can be removed by a few nights of trapping, using lights and fresh meat baits.

Herbivorous fish perform a special function; the Chinese say, 'if you feed one grass carp well, you feed three other fish'. Grass carp consume massive quantities of vegetation (their own weight in a day), and excrete large quantities of partially digested materials, which directly feed bottom-feeding fish such as common carp, and stimulate production in other parts of the food web. Grass carp can grow as much as 3–4 kilograms per annum. (The Chinese grow mulberries as the fruit is given to ducks and fish and the leaves feed shrimp and grass carp.)

Note that introduced exotic fish, mainly carp, have been enormously destructive in Australia's waterways. Take extreme care and seek advice before importing new aquatic species into your waterways.

Other animals

Yabbies like to hide in little things—they will dig holes in dam walls—and love living in beer cans (non-rust aluminium) suspended from a raft. Yabbies will eat worms. In marshes and wetlands make sinks to grow fish, yabbies and prawns. This can also be done in mangroves.

Freshwater mussels can be grown on ropes and will filter over 700 litres of water per day, cleansing like kidneys. They also deposit phosphate.

Good combined enterprises include:
- ducks and fish—in Hungary, 500–600 ducks are used per hectare
- fish and pigs—the basis of the traditional VAC (Vuon Ao Chuong) system in Vietnam
- fish and domestic waste

- fish and agricultural waste; for example, rice hulls and crayfish
- fish and industrial waste; for example, abattoirs, sugarbeet processing
- fish and worm farming.

A successful polyculture has a mixture of fish, crayfish, plants, molluscs, waterfowl and edge plants.

Management

Your pond will not remain crystal clear. Decaying plants, animal wastes, silt build-up and algal growth will cause the water to turn a pale green colour. This is a natural process and is an indication that the pond is functioning properly and contains sufficient nutrients to sustain the food chain. However, if the water is dark and murky, it can become toxic. This is more likely to happen in summer and fresh water must be frequently added to prevent this occurring.

A small amount of silt on the bottom of the pond is natural and is an important source of nutrients for plants and micro-organisms. In fact, the silt can be used as a source of compost for the rest of the garden.

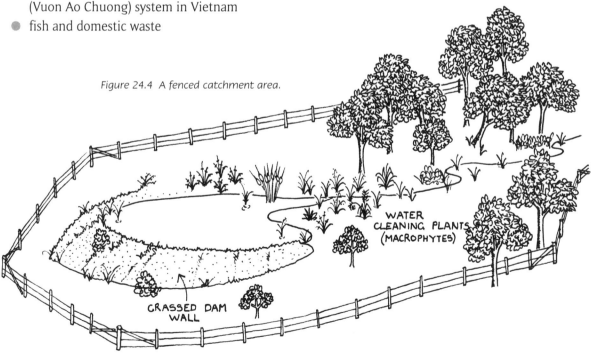

Figure 24.4 A fenced catchment area.

WATER CLEANING PLANTS (MACROPHYTES)

GRASSED DAM WALL

The requirements for management of small ponds are:

- a waterproof lining
- submerged oxygenating plants such as tape grass, water milfoil, water thyme, etc. and waterlilies on the surface
- scavengers to help establish the natural balance between fish and water snails and clean up rotting vegetation
- algae and goldfish to eat mosquito larvae and other insects
- mature water for the plants and fish, so do not empty the pond unnecessarily and top up the water gradually
- fertilising with small amounts of compost or manure.

All fish require oxygen and if there are too many fish confined in a small pond they may suffer from an oxygen deficiency. Pumps and water flows (channels of flowing water between ponds) can be designed to increase oxygen levels. Lack of oxygen can occur during hot weather because warm water contains less oxygen than cold water. It may also occur after rain when organic matter such as animal manures and vegetable matter has been washed into the dam. Decomposition of organic matter uses up the available oxygen. Signs of oxygen deficiency are dead fish or fish coming to the surface gasping for air. Oxygen can be replaced by circulating the water or pumping it up and spraying it back onto the surface. Ducks and geese are also very good at oxygenating water as they dive and swim through the water. Or you can take yourself for a swim.

Fish-eating birds will be attracted to a dam. Cormorants are the major bird predator. Their visits are infrequent and irregular, but once a dam with fish is found the birds will continue to work the dam until the majority of fish are taken. Fish quickly learn about safe retreats. Earthenware pipes, hollow logs, old tyres tied together and plastic pipes can be used, but not metal pipes or chicken wire as they release chemicals into the water. Thin fishing lines strung across the dam will deter birds without damaging them. Better still, create places for the water animals to hide.

On larger farms, manage your catchment area, dam and plantings as one organism. For example,

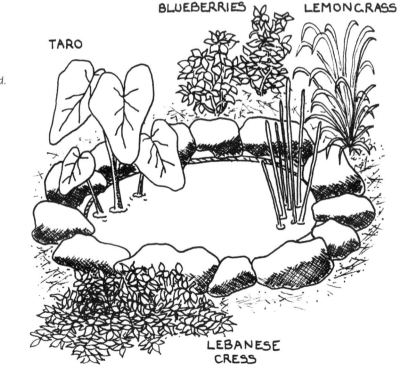

Figure 24.5 An established tyre pond.

TARO

BLUEBERRIES   LEMONGRASS

LEBANESE CRESS

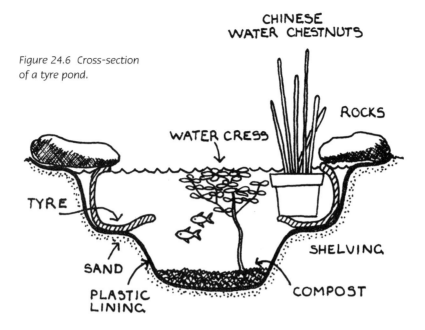

Figure 24.6 Cross-section of a tyre pond.

no chemical sprays should be used in the catchment area. Large stock should be excluded, and the immediate area around the pond should be protected by vegetation to reduce water muddiness.

Urban aquaculture

Aquacultures don't have to be large to be productive. Mini-aquacultures can be cultivated in old washbasins, baths and fish tanks. If you have sufficient room, tyre ponds can be a fine addition to a garden (see Figures 24.5 and 24.6). These small water systems can be placed throughout the garden. They will change the humidity and the light around the immediate area; they will help moderate temperatures; and they will attract beneficial insect predators. A small array of water-loving plants will grow well around them.

One duck will be quite comfortable in a tyre pond but two will be a little crowded. Rob has placed two small tyre ponds in his garden. As the water quality deteriorates in one pond, his ducks move on to the other pond. This means that Rob does not always have to be cleaning out the one pond.

 Try these:

1. Design an aquaculture system for your site and list how many functions it will add to your place. Integrate it with the household and farm water.

2. Develop a multiple-pond system and see how much diversity you can introduce. You should now have a whole site water plan.

3. List the plant and animal species you would introduce and the advantages and disadvantages of each. If you need more information you can talk to agencies that can help you, like the Department of Agriculture, the Department of Water Resources, your local fish market and fishing associations.

4. Design a tyre pond for a garden and list your reasons for:
   - site selection
   - choice of plant species
   - anticipated yields and advantages.

PART FIVE

Social permaculture

You have looked at the design ideas in permaculture and read and thought about disciplines in applied science, and then integrated them. You have the skills and knowledge to live well and simply in your bioregion. You have ethics and principles to apply to solving the problems that arise when living sustainably.

Of course that's only a beginning: you will have to research some details for yourself. You might have to find out the breeds of chickens that thrive in your environment or what fruit tree cultivars are best suited for your soils, but most of this information is easy to get.

The importance of working together for stability and security was covered in Part Four, where you saw how your design and land needs to be protected by co-operation with neighbours and similarly minded people. Now it is time to look at social permaculture.

Course participants have told me that for them this part is so important that it should be the first topic in permaculture. Social permaculture refers to the 'invisible structures' that we don't always see but whose impact we always feel. To take a recent example, cyclones in the US petrol belt caused oil prices to rise around the world. That is how invisible structures work. When interest rates rise 0.1 per cent around the world, the Malawi government doesn't have enough money to vaccinate children against smallpox because the World Bank ensures the government first pays off its foreign debt.

This part of permaculture reaches deeply into our lives. Bill Mollison was once addressing a group in London. At the end of his talk, someone asked, 'Bill, how do I build an organic vegie garden?' He replied, 'Read any one of 1000 books, but what I want to know is where do you bank your money and how do you spend it?'

Bill's answer cuts right to the heart of our society. This is what Part Five of the book deals with.

CHAPTER 25

Money matters: permaculture, wealth and livelihood

Understanding money, how it's earned and where it goes, is fundamental to good permaculture practices. Formal economics theory is frighteningly limited and dishonest about real costs. Until the late twentieth century, economics ignored the cost of producing goods, especially non-renewable resources. This is true for capitalism as well as communism. Every natural resource was 'free' and up for grabs, available to be exploited, often resulting in great human misery.

Fundamental to the preservation of natural non-renewable resources and our communities is how we earn and use money in our lives. Would you spend money directly to support enterprises that degrade the environment through pollutants, dangerous or excessive wastes, armaments and biocides or drugs? Would you support enterprises that exploit or harm people through discrimination, corruption, labour exploitation or unsafe and polluted workplaces? Of course not. Yet, that is what we are doing every day when we buy goods and services and deposit money in banks.

Banks have become very powerful and are tied into the global economy. Despite formidable evidence to the contrary, we deposit money with them because we believe that our money is safe, and that if we have enough money that we, in turn, are safe. We forget that when we put money into a bank we are lending it to the bank to do things that

we would never contemplate, yet it is with our money that the banks support destructive enterprises and practices.

Banks use and lend money for environmental degradation, suppress human rights, finance armaments and deplete the world's resources while polluting the atmosphere. Banks are ruthless about closure, especially on small businesses, and extract unacceptable levels of interest from loans and for the fulfilment of basic human needs like housing. The result of such power is that we abdicate our rights and voices. And yet, surely one basic human right is not to be in debt?

Financial advisers are reporting that investors of the future will want to know their returns are clean and green. There is growing global pressure for accountability and there has been a large swing towards investing in ethical projects. This is a worldwide movement and, because the research to

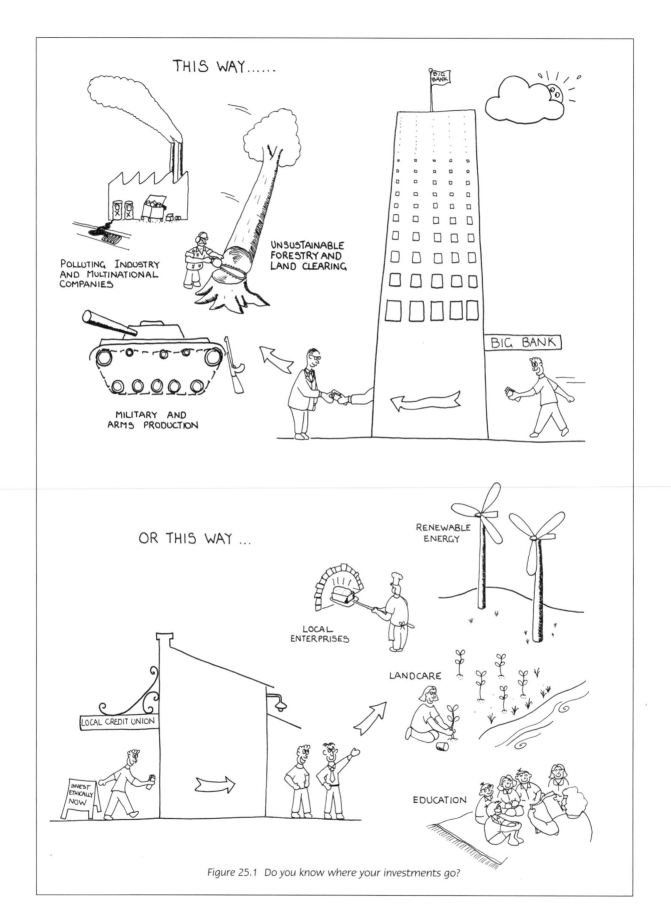

Figure 25.1 Do you know where your investments go?

ensure projects are clean and green is so thorough, the risks are somewhat lower than for the open market.

Our ethical task is to:

- acquire and spend money discriminately so that it does no harm and regenerates society and the environment.

Our design aims for money, wealth and work are to:

- live better with less
- live in safety and health using renewable local resources
- give back some of what we've used
- replace resources
- bring our ethics into line with our finances.

If we don't have aims for money, wealth and work:

- we will inadvertently support industries and products we would never deal with normally
- we will contribute to the exploitative use of resources.

Economics

The *old economics* is wealth based on cheap, natural resources, most of which are non-renewable and polluting. It also relies on cheap labour. This type of economics is now being questioned and challenged.

Ethical or *green economics* places economics in a matrix of interdependence with society, ecology, and ethics, with all human activities guided by ethical considerations. In practice this means that every enterprise is fully costed for its destructive as well as financial outcomes. A deforestation proposal would have to be costed for woodchips, soil and nutrient loss, river silting, erosion and habitat loss, air pollution etc. These are all costs that would either not be considered under old economics or which, worse still, would be subsidised by our taxes.

The honest assessment of present and future real costs leads to what's called 'cradle-to-grave' economics, or lifecycle cost economics. It applies to items such as staples or ballpoint pens or pencils— what is the real cost of surveying, mining, manufacturing, packaging and then disposing of one staple/pen/pencil? These distinctions become clear when you apply the cradle-to-grave test. Ballpoint pens and staples cost the Earth, whereas pencils are low cost.

Some economists are now advocating *zero growth*, and say that with recycling and new industries based on renewable resources we can accept a no-growth economy. There may be discomfort in the transition stage, however; thousands of small cultures lived in no-growth economies for centuries.

Start ethical economies in your bioregion

Rather than transforming the national and global economy, start local ethical economies where information moves quickly. People in bioregions, perhaps through permaculture design courses, can rewrite their lives and their use of money to include the values of Earth-care, people-care and the development of socially sensitive products. You become a 'non-buyer/non-consumer' of unethical banking, products, services and enterprises.

A lack of care, respect for people and ethical use of money is represented by dangerous foods and medicines, unsafe and polluted workplaces, addictive substances, labour exploitation and all forms of racial, religious and sexual discrimination and corrupt regimes. People practising bribery and price-fixing or gaining excessive profits and monopolies also represent a lack of ethics.

The multiplying power of money

Within communities, money multiplies goods and services as it stays and circulates in the community. One way to community prosperity is to keep as much money as possible circulating in the bioregion. This requires a return to local buying and selling of goods and services. For example, a measure of 'leakage', or loss from the community, is how much money the local supermarket sent out of the area last year compared to how much it spent on wages and local products. Where does the lost money end up—London, New York? Who does it benefit?

Disadvantaged people

An effective bioregion accepts responsibility for, and assists, self-reliance among marginalised or disadvantaged people. The effect of lifting poverty by starting at the base is that the welfare load is lightened. Trickle-down wealth almost never works because the rich:

- don't like to let go of their money
- do not put their money into people but into capital, wasteful consumer goods and speculation.

Act effectively and regain power

You do not have to be political to act effectively in your own life. Your power is available as soon as you use money and resources ethically. Your life will also be simpler. The great benefit of personal, life-affirming actions is that you are free and independent of political cant. Education and information are fundamental elements in turning yourself and society around. They are not difficult to arrange in bioregions and word of mouth is still mighty powerful despite the claims of mass media.

One example of this is in the Blue Mountains near Sydney, where I have been teaching permaculture since 1987. I started with six students. They spoke to others and from then on each course has had up to 20 enrolments. The courses were never advertised and I moved from the local community college to a neighbourhood centre. We soon had people with gardens and orchards as demonstration areas and we had people knowledgeable about bioregionalism.

One group of students started a local enterprise trading scheme (LETS), which grew to more than 500 members in about 18 months. As a result, the food co-operative, with its ethical purchase of quality food, was better supported. Another group started an organic gardening program on local radio, and others began a community gardens scheme. There is now a solid body of people with knowledge and skills who feel they live in an ethical bioregion. This has all been accomplished without advertising. The permaculture courses and LETS were not advertised, and yet most residents would know about one or both of these and each creates its own networks and links into other schemes and projects. We all feel better for living in a bioregion and knowing how to shape it in more sustainable ways.

Figure 25.2 Support your local markets.

Figure 25.3 How does your lifestyle contribute to the quality of your life?

Strategies for the ethical use of money

We can reject the idea that numbers written in a bankbook represent wealth. Instead, we look at other indicators of wealth. We take health, a safe society, peace, creative education and a clean sustainable environment as our ideas of wealth and we direct our money towards these ends. The latest social research has found that:

- up to a point, satisfying our needs and wants makes us happy, but after that we actually become unhappy and one in four people in rich countries is on antidepressants

- after a shopping spree, people return home depressed—the goods haven't met their emotional needs.

Spend to improve your quality of life, care for the Earth and people

An innovative program run by Richard and Maria Maguire increases your awareness of what you do with your money by asking you to keep precise records of your spending for one month and then analyse it. As you spend money you record three things and give each a mark out of ten.

1. How it contributed to your quality of life, health and wellbeing.

TABLE 25.1: SPEND TO IMPROVE YOUR QUALITY OF LIFE

Item	Cost $	Quality of life	Positive impact on environment	Positive social impact
Sunhat	15	9	5	3
Organic vegies	10	9	9	8
Whisky	8	5	4	6
Secondhand book	4	8	8	2

2. Its environmental impact; whether it's recyclable, durable, uses renewable resources.

3. Its social impact on labour and communities.

At the end of the month you add up each column and then cut out or make changes to those items you don't value. The bonus is that with this program you actually start to save money.

I found that I don't like sitting in noisy coffee shops drinking lukewarm coffee in small cups. I have became far more selective and prefer to take good coffee in a vacuum flask and sit in a quiet park and share it with a friend

Earning an income through right livelihood

Every bioregion is impoverished when people carry out activities that reduce their self-reliance and treat wealth as the getting of money not the meeting of needs, which includes income. Chilean economist Manfred Max-Neef has developed a list of fundamental human needs that are universal and unchanging. His list includes food, protection, affection, understanding, participation, creation, recreation, identity and freedom. Thus his concept of wealth becomes: income, health, quality and quantity of work, environmental quality, personal and social security and an individual's emotional and spiritual life.

The getting of such wealth to meet these needs depends on time and culture, so there is room for creativity and diversity in earning livelihoods that do not damage society in any way. This is the concept of *right livelihood*. It is not a new concept because there have always been cultures and religions which have dissuaded people from exploitative work. For example, for 400 years Quakers have practised right livelihood by not accepting employment in any field which is destructive to society. That ruled out tobacco, banking, alcohol, defence and so on. However, the principle of right livelihood has gained impetus because socially and environmentally destructive work now threatens us and the future of life on Earth.

Human labour is a renewable resource and there is plenty to do. People can act as curators of their

bioregion and from this will spring a multitude of work options. There are those associated with shelter and technologies; those connected with food, its growing, gathering and processing; with administration and finance; with environmental care; with education; with arts and leisure.

Reclassify your assets

Reclassify your assets to see which ones lead to better health, water and food security compared with those which make us more vulnerable.

- *Degenerative:* These deplete non-renewable resources and they decay, rust, wear out or pollute. They include buildings, cars, machines, computers, dams and powerhouses. Governments of all sizes tend to spend on these assets and call them infrastructure. Banks mostly lend for degenerative assets.

- *Generative:* These process or manufacture raw materials into useful products. Examples include grinders, blenders and lathes.

● *Procreative:* These multiply over time, such as fruit trees and hens.

● *Informational:* This includes ethical knowledge and skills leading to self-reliance.

What we can do

● We can choose where we will put our money.
● We can buy only ethical goods and services, which makes shopping much easier.
● We can lend money to each other at zero or no interest.
● We can lend our money to socially responsible investment (SRI), also known as ethical investment, which supports co-operative housing, local integrated transport systems, afforestation, clean food and processing, and making durable useful and necessary products.

Four types of investment

There are four categories of investment, which determine whether our money is used creatively or destructively for our society, and ultimately ourselves.

● *Active:* Invest money and work in sustainable projects, such as afforestation, local organic nursery, co-operative housing and local employment.
● *Passive:* Buy products from ethical companies.
● *Neutral:* Spend surplus money on arts and leisure.
● *Unethical:* Assists in retailing of dangerous, destructive goods and services.

We should boycott unethical investments, even if only out of self-interest, because they will eventually rebound on us. We choose neutral investments when we have too much money or when all our other needs are met.

Socially responsible investment (SRI)

Ethical investment companies or socially responsible investment actively diverts money from destructive to creative and sustainable uses. People with some surplus of income place it so it works responsibly. There are now many companies to assist you with ethical investment. Some churches are placing their funds in them. Quakers in Australia recently invested all their yearly meeting funds ethically. This was the wish of all members. The Uniting Church of Australia has an ethical investment portfolio and there are many others.

If you want to invest responsibly, there are three ways to do it:

● *Divest:* Find out what your bank or financial company is doing, and if you don't like it, take your money out of it. For example, you may find they support armaments, logging tropical rainforests or the nuclear industry.
● *Affirm:* Invest in those things that are socially responsible, such as community housing, jobs and land rehabilitation.
● *Target:* Invest in a company by buying up shares until you have sufficient majority to vote at a shareholders' meeting and change their policies. This was done with Volkswagen, which was clearing rainforest in the Amazon. The company's policy was reversed.

Alternative local financial bodies

It is important not to have an open 'money artery' which drains money away from your bioregion. One objective of a bioregion is to have 40 per cent of all trading in local dollars and organisations. Apart from its value in local communication, education and full use of human potential and goods, this 40 per cent builds local wealth. There is a huge potential market for it in any bioregion. It enables the community to:

- avoid multinationals
- support local enterprises
- support local currencies
- meet basic needs for goods and services.

A *Local Enterprise Trading Scheme* (LETS) is one way to achieve this. It makes use of a local currency, also known as green dollars, which you trade with in your bioregion, and the LETS circulates a newsletter with offers and requests of goods and services.

Revolving Loan Funds, in which people put in money and then lend it to members or people in their bioregion, have been very effective in some places—among them Sri Lanka (Thrift Societies), Bangladesh and Vietnam. In Australia, the Bellingen Loan Fund in New South Wales will accept deposits from outside the region but only lends inside the

region. In developing countries, and probably others, they work well when they are a saving fund. Those which give only credit work far less well and often only serve to take people into debt—as the banks do.

Outside your bioregion

Invest in helping people in poorer countries with appropriate technologies or to retain valuable natural forests and build environmentally sound houses.

 Try these:

1. Make a list of your assets according to the four types: degenerative, generative, procreative and informational. How rich are you according to this reclassification? How many assets are durable and environmentally friendly?
2. Describe the type of socially responsible investments you would like to see in your bioregion. Outline the type of organisation that would carry it out.
3. What organisations do you belong to which help to hold money in your bioregion?
4. Find out about ethical investment companies and arrange to put your superannuation in one.

CHAPTER 26

A sense of place

Social permaculture is all about how we live in our villages, towns and neighbourhoods. We may be citizens of a large nation but we do not 'live' in it daily. In reality, our education, business and leisure are carried out in limited familiar places and on a 'people scale'; that is, within small personal groups at home and at work.

When I was teaching in Vietnam there were 25 people in the class. We were talking about bioregions and I realised that my interpreter, Gia, was frequently using the phrase 'my country'. For a while I thought this meant Vietnam, except it reminded me of the Aboriginal phrase 'my country', and so I asked each person what the words meant for them. Every person explained that their 'country' was where they had been born and where they grew up. Each valley, each stretch of land between delta canals, each delta island or each coastal strip in Vietnam had a clear and unambiguous identity for the person who had been born there. I asked them to tell me what 'their country' was like. Their enthusiasm rose as they described songs and dances, special fruits like mango varieties, housing and accents. They went on to speak of the war and how women and men went to fight together from 'their country', and when bombing got bad they sat in bunkers and the jungle telling the stories and singing the songs and remembering the special foods of 'their country'.

Then they asked me where my country was. I explained that I did not have a 'country' in that sense, although I have lived in many places and now live in the mountains outside Sydney. They found it difficult to understand how I did not belong to some land, 'my country', where my ancestors' bones rest.

To some extent, however, permaculture makes use of a similar concept—that of the bioregion. While the term has been mentioned several times before in this book, it is now time to look at what it means in more detail. A bioregion is an association of people who live in a natural and definable region. It can be defined by roads or water, language or trees. For example, my bioregion is partly defined by the limit of growth of the *Angophora costata*—the Sydney red gum—which grows sparsely at altitudes greater than 750 metres. The acid test of a bioregion is that it is recognised as such by its inhabitants. There are differences in people's abilities to define their bioregion. People living in the Southern Highlands of New South Wales are absolutely clear about their boundary and residents of the Pittwater area of Sydney have no trouble defining their bioregion. People living in the suburban sprawl of large cities have more difficulty, however.

Bioregions are people-sized in scale. They vary considerably in available resources, climate and size but they can meet the needs of the residents from within the area and within reasonable distance. Your land can be made more sustainable but the true essence of sustainability lies in a bioregion. A bioregion can be self-sufficient whereas an individual cannot without having such a hard life that they die young of overwork. The real success of a bioregion depends on the way people work and it fails unless societies move towards co-operation as the prevailing mode of interaction and communication. Bioregions also need political and financial units.

The ethical task of a bioregion is to:

- preserve and develop our natural character
- improve our sustainability
- develop biological and human resources which are our true wealth
- care for our resources and build self-reliance.

The design aims for a bioregion are to:

- give priority to developing and maintaining bioregional stability in resources, work, finances, etc.
- keep trade within the bioregion
- develop fast communications systems
- ensure that everyone in the bioregion has one or more links to a bioregional organisation.

If we don't have aims for a bioregion:

- our money flows out of the bioregion
- we are flooded with imports
- we have more unemployment and migration out of the area
- we are less able to withstand adversity.

The importance of a bioregional directory

Bioregional stability is measured by a reduction in imports and exports and this means an associated growth in self-reliance—the bioregion moves closer and closer to meeting its own needs. A valuable tool for achieving this is a bioregional directory. This is a guide to, and a benchmark inventory of, all the services and resources that have ethics compatible with those of the bioregion.

The directory takes an inventory of resources as shown in Table 26.1. This list can be continued for education, work (right livelihood), finance, trade and so on.

The directory achieves three things:

1. It is an ethical buyer's guide for people who want to support the bioregion.
2. It shows what openings are available for people to produce products, find a work niche or train in a specialised area.
3. And finally, when regularly produced, it tells residents to what extent they are meeting their own needs (becoming self-reliant) as the 'gaps' list decreases.

Every resource, service or product has to meet given criteria. For example, criteria for food production are:

- locally produced and processed
- exotic and indigenous species
- organic and free of biocides
- good nutritional value.

Criteria for shelter and buildings are:

- local sustainable materials
- local labour

TABLE 26.1: BIOREGIONAL DIRECTORY

Food		Shelter	
Available	Absent	Available	Absent
teas	coffee	mudbricks	tin
honey sugar	beet sugar	straw bales	tiles
		stone	carpets
		rainwater	

Figure 26.1 A stable bioregional centre where ecology, economy and culture are interwoven.

- energy- and water-efficient designs
- non-toxic materials and furnishings.

The directory can also offer non-formal education through short training courses. For example, under the heading 'Shelter and Buildings', plumbers can find refresher courses on water-saving devices and biological water cleaning, and how to make whole site water plans.

Bioregional prosperity

Bioregional wealth is measured by the increase in biological resources. The growth in plant and animal diversity, in community gardens and in urban forests are all indicators of regional wealth. However, it can also be measured by the development of people in skills, jobs and ability to work co-operatively. Other indicators are a decrease in domestic violence, stealing and need for police.

Three interwoven issues in a bioregion are control of the land, the economic system and people's livelihoods. Each of these helps the

bioregion to become more sustainable and autonomous while staying within local control. In Chapter 25 we looked at the principles of green economics and right livelihood. We will now consider the issues surrounding land control.

Control of land

Permaculture sees land as a resource to be used for shelter and food, and not a commodity to buy and sell. Most land is degraded by the people who buy it: it is chemically polluted, grassed, mowed, concreted and fenced. Both capitalist and communist models of land ownership have been disastrous because they regard land as an inert commodity from which to extract maximum productivity with minimum inputs while paying no attention to long-term effects.

It is an irony that the very people who cared for whole ecosystems in our bioregional sense, the indigenous people of the world, were dispossessed of their land by people who believed that tribal people did not use land well.

All bioregions should include in their ethics the reinstatement of land to indigenous people, and priority of access to unemployed people and marginalised people such as the young, elderly, new migrants and refugees for shelter and to grow their own food. Those who need it would be offered education to design the land.

- In Vietnam when you need land for housing and to grow food you go to your local People's Committee which allocates you an appropriate parcel.
- In Botswana people know which land is their father's land by the white stones. They have access to this land.
- In Australia in most places, you must find more than $100,000 to purchase land and then spend the rest of your working life paying off the bank mortgage at immoral interest rates.

Access to land for shelter and food is really a natural right, as is the right not to be in debt to banks. However, access to land is associated with the responsibility to leave the land better off than when it was acquired, which means more sustainable and richer in biological resources. It does not mean treating land as a commodity for financial gain.

In many cases the cost of land does not reflect what the land will produce. Realistic land tenure would be to lease 'right of use' to individuals or communities and the bioregion would have an ethical charter to prevent land abuse. It would act as a custodian to monitor activities and resources.

Land access opportunities

If you live in an area where it is difficult to acquire land for basic needs, there are several models for you to try. Your bioregional office can set up a Land Access centre or desk that is tailored to your region and offers people, starting with those with the greatest need, the following possibilities:

- *Oxfam model:* The office co-ordinates those who want land to grow food and those who have land and are willing for people to use it, such as local councils and elderly people, hospitals and schools. The office works out an annual renewable lease agreement between the two parties.
- *City farms:* Interested people in the bioregion negotiate a lease with council (never for less than five years) for public land close to the town centre and transport. Some of the city farm's activities will include: nursery, worm farm, tool rental, recycling centre, demonstration, allotments, domestic animals, family–community meeting rooms, picnic space, seeds, resale items and a classroom for teaching people food-growing skills. It can later open to visitors and schools.
- *City as a farm:* This is when you or a group decides to harvest city surplus products;

Figure 26.2 Making use of vacant land.

for example, chestnuts, figs, mature trees for furniture or firewood, grass clippings for mulching, or whatever is available.

- *Farm Link:* This is also known as Community Supported Farming (CSF). It is a producer–consumer co-operative where a number of town families contract a farmer to grow food and products to their requirements. At quarterly meetings with the farmer, everyone decides what will be needed for the next season. Families spend some time each year working and helping with planting, harvesting and processing.

- *Farm and garden clubs:* A group buys a farm and either the farmer stays or another person manages it. Extra accommodation is built for the new owners and the farm may have special enterprises, such as aquaculture.

- *Commonworks:* This is a farm held by a Land Trust that is close to a town or city. It arranges a series of special leases on the land for forestry, livestock, crafts, teaching, a nursery and even mudbrick workshops. Ten per cent of the income is paid back into the Commonworks Fund, and this pays the rates and maintenance. In Vietnam the local People's Committee gives land to groups of disadvantaged people (there is no social security), such as elderly people with no families, handicapped people or orphans. The people can live there and, in return for work, have enough to eat and sometimes sell surplus at the local markets.

- *Government land:* Many government departments hold land that they are unable to maintain satisfactorily, so that they become infested by feral animals or weeds. A community group approaches them for CCM—that is, Care, Control and Management of a piece of land for, say, food gardens. This is good use of land and good advertising for the department. Departments that hold land are Railways, Education, Main Roads, Planning and Crown Lands. So do schools, churches and hospitals.

- *Advertising:* Advertise in local papers and LETS magazines to care for land for others. Some people are delighted to have their land looked after.

- *Land restored:* In 1988, the year of Australia's bicentenary of European colonisation, the Quakers in Queensland gave half their total land to the Aboriginal people as a gesture of reconciliation.

- *Buy for others:* There are now proposals to buy land and secure it in a trust for people excluded from land to satisfy their own needs. For example, community gardens, local woodlots, and aquaculture systems could be dedicated as Commons always accessible to disadvantaged people.

Land use by individuals and groups

The success of bioregional land use is measured by the extent to which it meets human needs and stays in harmony with the natural environment. Write your own ethics for land in your custody. Land is only entrusted to you for your lifetime so write ethics for the future. Some ethics for you to consider include:

- to develop its resources wisely
- to leave it better off than you found it
- to increase the biological diversity
- to act in ways that increases its sustainability
- to introduce no polluting materials onto it.

 Try these:

1. Join your local community gardens and teach others to grow food.
2. Think of a subject you've always wanted to learn, find some other people and then hire your own teacher.
3. Hold a street party, which can be a meal, a children's party, a 'get to know each other' party or simply drinks before dinner.
4. Write down what you consider to be the limits of your bioregion. Work with friends and make a bioregional directory. Write criteria for services and products.

CHAPTER 27

Living in communities, villages, suburbs and cities

This chapter brings us home, to the place for which we have done a design. We need permaculture designs for more than just each family. Whether we live in a village, suburb or city, most of us live within networks of people.

Many wish they had deeper links to a community, as modern society is individualistic, insecure and fractured. Many people don't know where to start to lead different lives. In permaculture it is possible to build communities where we live.

To begin it is useful to have a sense of connection and loyalty to a bioregion. This is where you invest your energy and time and it is worthwhile because it is rewarded. This chapter is about building that sense of belonging wherever you live and looks at some possibilities and how they can offer more sustainable and richer lives by:

- providing mutual support and encouragement
- integrating family, work and leisure
- widening the possibilities to participate and have fun
- reducing the use of non-renewable resources.

Most of the world's people live in communities or villages spaced throughout bioregions. In the Third World many people who once lived in closely knit villages now live a degraded existence on the edges of cities, usually because they have been dispossessed of ancestral lands by the economic system, modern agriculture or colonisers. In the richer nations many people leave cities to find better lives in country areas and in communities. Most of these people are neither socially nor economically prepared to live in communities.

When people leave traditional communities, or when new communities fail, it is usually because of reasons associated with people or the land.

 ### Our ethical task is to:
- live co-operatively, thoughtfully and simply with each other.

 ### Our design aims for living together are to:
- conserve natural resources
- implement fair trade and financial systems
- share responsibility for work
- provide meaningful, dignified and safe work.

 ### If we don't have design aims for living together:
- we destroy our resources and live more degraded lives
- we have inequitable power and management hierarchies

- our projects will fail
- we threaten the future for our children
- we create divided communities.

Why communities succeed or fail

Many people dream of living in a harmonious community in a beautiful environment. In Western societies most people have never lived in a community and they have also been encouraged to be highly individualistic, so they need to learn new skills for living together with others. There is considerable research on communities and particularly why they succeed or fail. In general, it takes about ten years for a community to 'settle' and so it can be worthwhile to wait out the growing pains if you can see issues being settled.

Why communities fail

- *Cheap marginal land:* The crops fail, or were unrealistic, or cost too much to develop; for example, putting in dams and windbreaks to provide the necessary water security and protection.
- *No capital for roads, dams and fences:* The farming systems fail or people borrow money from banks and moneylenders who eventually repossess the land.
- *Lack of realistic objectives:* Not everyone will get out of bed on Sunday mornings to fix the holes in the road.
- *Ill-defined ownership:* Personal and property boundaries are ignored; there is resentment. Sometimes settlers do not have legal rights in the eyes of local authorities and can be thrown off the land.
- *No agreed design:* People agree in the beginning to 'care for the land' then someone brings 200 goats and others grow irate. An agreed design settles the question of what enterprises can be carried out and where, who will harvest and when.
- *No forward planning:* This can result in no money for rates, water, fences and other costs. Sometimes it can lead to failure of crops due to wrong planting times. It is called ad-hocery!
- *Isolated communities or difficult to farm:* People get lonely, too tired and want to leave.
- *No framework for decision-making:* Disagreements break up the community.

Why communities succeed

Communities succeed if the above factors have been resolved and if there is consensus on the ethics. Meeting procedures need to be designed to ensure conflict resolution and to move stalemates forward. The first meeting is held to agree on the ethics.

The only meetings after the ethics are agreed upon are for work. In this case groups of volunteers with motivation and skills roster themselves to carry out an important job. Discussions happen at work—on the job. Others in the community must leave them to do the work and not interfere. This requires *trust*. People must be trusted in their areas of expertise. So, people are trusted equally to dig holes and paint walls as they are to do the accounting, childcare or electrical work. People usually trust others as much as they trust themselves and their own ability to be skilful or responsible.

In public discussion the questions *who* and *why* are not asked since these are adversarial and cause people to go round and round in circles breeding intolerance and grievance mentalities. Public questions are *how* and *when*. These carry the objectives forward. (Privately, over fences and in chats while working, people will of course ask 'who' and 'why'. For example, 'Why did you break the fence?' will be asked in private but becomes, 'How can we fix the fence?' in public.)

Limits are established on the number of people the land can support at various levels of consumption and waste. For example, how many people can an area support when they want meat every day, or want to use 1000 litres of water on their lawns? How much sewage can land absorb? How many trees have to be cut for everyone to have wood fires?

Residential ownership of the land should be an absolute minimum of 60 per cent, and 80 per cent is far better.

Levels of public and personal *privacy* must be defined and clarified.

Community management

Proper management of a community is very important and there are some inspiring models to emulate, drawn mainly from small populations of indigenous people. These seldom rely on authoritarian models of hierarchy, which result in injustice. Nor do they advocate majority rule in which important ideas and the feelings of minorities are neglected, usually resulting in divide-and-rule. This does not imply anarchy, however, which can result in alienation, confrontation, and lack of empathy and communication.

Consensus decision-making can result in stagnation if one person boycotts a decision and withholds consent as a means of exerting power and control. Quakers have a 500-year tradition of consensus, which has worked very effectively for extraordinarily difficult issues. They also have an agreed process to achieve consensus, which entails silent and thoughtful consideration of issues after listening to others and before speaking.

Hierarchy of functions for working communities

Communities work well where there are no hierarchies of decision-making but there are hierarchies of function or work: where everyone agrees that some issues are very important and some more important than others. Once this is done, individuals or small groups take responsibility to do their jobs and do them well. Examples of hierarchical functions are:

- the care and safety of children
- keeping the access road open
- maintaining clean and sufficient water for essential needs
- running a newsletter
- food supply and preparation.

Financial management, although very important, is not the first in the hierarchy of functions of a sustainable community.

Permaculture designs for villages and communities

Global Ecological Network (GEN) is a worldwide eco-village network. There are now many successful communities, and governments in Germany and Denmark see them as the new models for clean, safe and economical living.

The problems of small rural settlements such as villages and communities are more often caused by world global recession and the subsequent struggle to survive. These communities often lack information, education and skills. They need people who can assist with ideas for LETS and work groups, who can help with alternative enterprises and sustainable and self-sufficient living. Many small rural towns are becoming sad places where people blame themselves for not being able to cope. However, there are places, like Rylstone in New South Wales, which has welcomed new people, started festivals and encouraged new ideas. The town has turned its economy around.

Residents with permaculture knowledge

It would be ideal if every resident completed a permaculture design course. However, if this is not possible then ensuring that 30 per cent or more of the residents have permaculture knowledge should provide the skills to enable the planning to be done. Again, when designing communities and villages, use the ideas of zones.

The design

The physical layout of villages and communities is similar to that of permaculture farms and important issues must be decided early. Villages and communes are zoned 0 to V, and special protection areas—rivers, streams, dams, forests, ridges, wildlife and soils—are designed as inalienable. They require tight ecological controls. The next constraints are

Figure 27.1 A suburb as a cluster of sustainable living.

on homes so that they are non-polluting, and energy and resource conserving. Then decide how much land will be held privately and how much in common.

Communities use the same principles of energy expenditure and distance to designate zones. In this case, however, orchards can form corridors between rows and houses, and Zones IV and V are the encircling commons with water-harvesting dams. Decaying villages being renewed can be retrofitted along these lines as well.

- Zone I is home and food gardens.
- Zone II is orchards, footpaths, and close open space like playgrounds.
- Zone III is larger open space and community gardens—often staple crops.
- Zone IV is the reserves, fuel forests, nut forests and windbreaks.
- Zone V is planted for wildlife corridors, native plants and sanctuaries and for providing a framework for the other zones.

Always make sure you have your disaster plans well prepared and rehearse them regularly about three or four times a year.

Permaculture designs for suburbs

Suburbs resemble car parks: they produce almost nothing and require huge inputs of goods and services. They are vulnerable to breakdowns in transport, food supplies, water and energy. However, they also have immense potential to be developed as clusters of sustainable living.

Neighbourhood clusters

This is the finest strategy for suburbs. Start in your street. Get to know people by having a street dinner. Let people know that their children are safe to visit you. Give plants to each other and talk over the fences. Then remove the fences. Begin working together and start small local classes to satisfy primary needs for socialising because of the many isolated people in suburbs.

Harvest human potential

Work in groups to plan streets and care for children, then to identify community needs and wants. Once

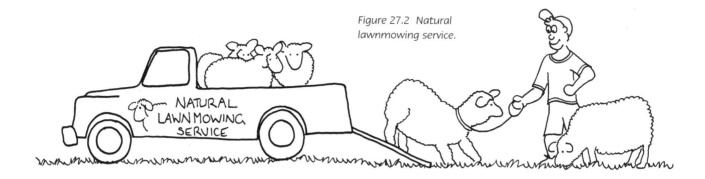

Figure 27.2 Natural lawnmowing service.

people realise how powerful self-reliance is they will think of converting car parks, schools, railway lines and ugly road verges into productive parklands.

Encourage income generation

Enhance the wealth of the suburb through teaching, learning, doing and marketing small clean industries such as fruit, nuts, honey and poultry. Bring sheep and goats in as lawnmowers and pets.

Think local

Local economies run best by buying, selling and trading within neighbourhoods, using both the official currency and green dollars.

Start small

Don't start too big or try to go too fast. People must be able to join in and get the feel of developing their place. Encourage a sense of belonging and responsibility. Later there will be markets, meadow lawns and special restaurants. Don't be too intimidated by restrictive bureaucracies; you don't need them for most things.

Help people towards self-reliance and confidence in meeting their own needs and work in small groups for disaster evasion or avoidance. For example, if wildfires come can the community manage if emergency services are elsewhere?

Permaculture in cities and large towns

Often people feel strongly about cities. They either love them or hate them. Whatever your feelings, there's no denying that cities do have problems.

Cities are congested and polluted and extraordinarily vulnerable. Local government increasingly does not have the money to clean them up. New York City, where more water is lost through leaks in old pipes than is actually used, hasn't enough budgeted money to service the repairs required for the water mains. New York's largest exports are sewage and garbage.

A city is vulnerable to strikes and shortages. Most cities carry enough fuel and food for all their people for about three days. Cities clog up with transport systems and drain resources away from rural areas. These resources (food, wood, fibre, dyes, fabrics, rubber, etc.) are transformed into waste, which has then to be exported from the city.

Self-generating economies, small businesses, and buying and selling in small areas will all help to regenerate cities. Incorporating agriculture and wilderness into cities is also essential. They contribute to its ecological balance and offer solid social, economic and psychological balance as well.

- Oregon in the United Sates requires every town to draw up and implement greenbelts and limit or prohibit city growth.
- Hong Kong grows 45 per cent of its own vegetables.
- New York found one-quarter of the city had 1000 pieces of spare land that could be used for gardens.

TABLE 27.1: THE AWFULNESS OF CITIES AND TOWNS

Trait	Elaboration
Light quality	This is poor with dark, cold shadows and glare from skyscrapers. Many buildings are empty due to high rents or dereliction. Cities are defined by an unnatural ratio of hard (concrete, bitumen) to soft (trees, water) textures and by angles and subsequent wind patterns called 'canyon effects'.
Temperatures	These are elevated as heat radiates from roads and buildings.
Pollution	Water pollution runs into creeks and rivers. In Sydney, Australia, 20,000 tonnes of pollution per day is pumped into the air from cars. One of the newer industrialised city diseases is particulate matter in the lungs from car and factory emissions. Asbestos from car brakes and dung from animals run into waterways.
Bare areas and vermin	Ground areas lack plant materials and attract many scavenging animals such as rats, cockroaches, seagulls and pigeons.
High dependence	Cities depend on piped, imported chemical water and, increasingly, bottled spring water.
Import monsters	They are huge consumers of fossil fuels, motor vehicles and electricity and this increasingly makes them super vulnerable to collapse.

Planting priorities

Plant to increase biomass and moderate dust, heat, cold and noise. Grow plants on cool and hot walls. Plants deflect noise, mellow bounce-back noise and insulate buildings. They take an edge off severe winds, and collect and filter pollution and dust. Pseudo-monas, a micro-organism which lives on leaves, filters positive ions and helps to create clean air.

Make urban forests along freeways, highways, disused power stations, railway land, car parks, schools, hospitals, nursing homes and playgrounds. Start with very dense plantings which are thinned out after three or four years and mulched for sale to gardeners, used for firewood or, if they are big enough, used for fencing materials and poles. It is necessary to start with pollution-resistant street trees that are proven survivors in your city. Avoid planting toxic or allergenic trees or those that drop squashy fruits. Nuts, seeds or oils can be harvested from trees planted in quieter streets.

Food in cities

People can start garden clubs and plant on balconies, rooftops and in windowboxes. Rooftop gardens are particularly good for stress management and relaxation because they give a feeling of space. Housing co-operatives and food co-operatives teach people to grow food.

Many annual plants with surface fibrous root systems grow well in cities and help absorb much organic waste. Food plants that work well in cities are bananas, pawpaw, tomatoes, citrus in tubs, cucumbers, beans, peas, passionfruit, chokos and chillies. In apartment housing people have traditionally grown pumpkins, citrus, figs, grapes, chokos and bananas. Grow fruit along backyard fences and on quiet streets to create an orchard park.

Figure 27.3 Many plants grow well in rooftop gardens and absorb much organic waste.

In suitable climates, macadamias, avocados, mangoes, figs, mulberries and pecans grow well in cities. Mix firewood and flowering trees with food trees.

What you can grow is limited only by your ability and imagination. There are many good books and websites now for growing food in cities (see References).

Community gardens

Orchards and urban forests start with Open Space action groups. In New York there are Green Guerrillas who plant trees after digging holes in the footpaths. New York has more than 1000 community gardens.

Retrofit buildings

Margit Kennedy in Berlin designed and retrofitted older city buildings and turned them into energy-efficient apartments. Today glasshouses combine with restaurants, solar hot water systems are fitted and, in one building, waste water is completely recycled to roof gardens. Ground floors, parking stations and cellars can be used for housing chickens and hares, which are allowed out to graze in gardens. Waste recycling of metals, glass, paper and organic matter is almost complete.

Transport of people and goods

- Let roads deteriorate. Design and implement clean public transport.
- Calm traffic by reducing the width of streets. Tax private cars as they enter the city and design streets so it is easier to walk than it is to drive. (Singapore has reduced its congestion and pollution by taxing vehicles entering the city centre, and providing superb public transport.)
- Give cycleways priority over roads. Use canals and waterways for the transport of major goods.
- Use electric railways to carry dangerous and hazardous waste.

 ## Try these:

1. Write about your neighbourhood, village or community—what you like about it, and what you don't.
2. Design some physical improvements that you would like to see.
3. Think of some community organisations or links that would make it stronger.
4. Do you feel that you belong there? If yes, list why, and if no, list what needs to change.

References, websites and resources

Although the references and websites included here are weighted towards Australia, they will take you all over the world. Of course, the fact they are here does not mean that I endorse everything in them.

Important General References
Carson, R., *Silent Spring*, Penguin, England, 1963
Dillard, A., Pilgrim at Tinker Creek, Picador, London, 1976
Earth Garden Magazine (www.earthgarden.com.au)
Flannery, T., *The Future Eaters*, Text Books, Melbourne, Australia
Fukuoka, M., *One Straw Revolution*, Rodale Press, Emmaus, USA, 1984
King, F.H., *Farmers of Forty Centuries*, Rodale Press, Emmaus, USA, 1988
Lowe, I., *A Big Fix: Radical Solutions for Australia's Environmental Crisis*, Black Inc Publications, 2005
Trainer, T., *Developed to Death, Rethinking Third Word Development*, Green Print, 1989
Wright, R., *The Short History of Progress*, Text Publishing, Melbourne, 2004

Introduction and Chapter 1
Books and magazines
Holmgren, D., *Permaculture: Principles and Pathways Beyond Sustainability*, Holmgren Design Services (www.holmgren.com.au)
Mollison, B., *Permaculture: A Designers' Manual*, Tagari Publications, Tasmania, 1988 (www.tagari.com)
Mollison, B. & Holmgren, D., *Permaculture One*, Tagari, Tasmania, 1978
Mollison, B. & Slay, R.M., *Introduction to Permaculture*, Tagari Publishers, Australia, 1991
Permaculture Magazine (UK) (www.permaculture.org)
The Permaculture Activist (US journal) (www.permacultureactivist.net)
Wilson, E.O. (ed), *Biodiversity*, National Academy Press, Washington, USA, 1988
Websites
www.acfonline.org.au
www.ecologicalsolutions.com.au
www.ibiblio.org
www.permacultura.org
www.permacultureinternational.org
www.permaculturevisions.com
www.permaculture-europe.org

Chapter 2
Books
Suzuki, D., *Genethics: The Clash Between the New Genetics and Human Values*, Harvard University Press, Cambridge Mass., 1989
Thompson, W., *Gaia: A Way of Knowing, Implications of the New Biology*, Ed. Irwin, Lindisfarne Press, 1987

Websites
www.davidsuzuki.org
www.earthcharter.org/news
www.ethics.org.au
www.holmgren.com.au
www.jaring.my/just
www.meaning.org
www.truecosteconomics.org
www.wme.com.au

Chapter 3
Books and magazines
Benyus, J.M., *Biomimicry*, Perennial, New York, 2002
Colinvaux, P., *Why Big Animals are Rare*, Penguin, London, 1980
Davies, P., *The Cosmic Blueprint*, Penguin, Australia, 1995
Ecologist, the world's oldest environmental magazine
Habitat, quarterly magazine of the Australian Conservation Foundation (www.acfonline.org.au)
Kauffman, S., *At Home in the Universe: The Search for Laws of Self-Organization and Complexity*, Penguin, 1996
Lovelock, J., *The Gaia Hypothesis*, OUP, Oxford, 1987
Recher, H., Lunney, D. & Dunn, I., *A Natural Legacy: Ecology in Australia*, Pergamon Press, Sydney, 1990
Suzuki, D. & Dressel, H., *From Naked Ape to Superspecies*, Greystone Books, Vancouver, Canada, 2005
Whitrow, G.J., *Time in History*, OUP, 1988
Websites
www.ecologicalsolutions.com.au
www.sustain.org/chain
Footprint analysis:
www.footprintnetwork.org
www.redefiningprogress.org
Footprint calculators:
www.earthday.net/footprint
www.myfootprint.org

Chapter 4
Books
Alexandersson, O., *Living Water*, Gateway Books, 1996
Boden, R., *Water Supply: Our Impact on the Planet*, Hodder, Wayland, 2002
Ghasseni, F., et al., *Salinisation of Land and Water Resources*, Centre for Resource and Environmental Studies, Australian National University, Feb 1995
Gleick, P., *The World's Water 2000-2001*, Island Press, 2000
Macdonald Holmes, J., 'The Geographical and Topographical Basis of Keyline', internet article (www.yeomansplow.com.au/basis-of-keyline.htm)
Singer, A., 'The Goddess and the Computer', *The Power of Water*, New York University Press, Ithaca, 1992
Yeomans, A., *Water for Every Farm: Yeoman's Keyline Design* (www.keyline.com.au/form03.htm)
Websites
Statistics for world use (and abuse) of water:
www.ata.org.au
www.citizen.org/cmep/water
www.clearwatertechnology.com.au

www.echotech.org
www.greenglobes21.com
www.ifpri.org/media/water
www.ofm-jpic.org/agua
www.polarisinstitute.org
www.postcarbon.org
www.reclaimedwater.com.au
www.unesco.org/water/wwap/wwdr/ex_summary
www.urbanwater.info
www.wateruse.org
www.wateruseitwisely.com
www.worldwater.org
www.wwrf.org
Water and farming:
www.nsfarming.com
www.nyserda.org/publications/solarpumpingguide.pdf
www.p2pays.org/ref/08/07682.pdf

Resources
Greywater Reuse Systems (www.greywaterreuse.com.au)

Chapter 5
Books
Carefoot, G.L. & Sprott, E.R., *Famine on the Wind*, Angus & Robertson, Sydney, 1969
Flannery, T., *The Weather Makers: The History and Future Impact of Climate Change*, Reed New Holland, Sydney, 2005
Linacre, E., *Climate Data and Resources*, Routledge, New York, 1992
Lowe, I., *Living in the Hothouse: How global warming affects Australia*, Scribe Publications, 2005
Websites
www.deh.gov.au/minister/ps/2006/psmr04jan06.html
www.greenhouse.gov.au
www.yeomansplow.com.au

Chapter 6
Books
Conway, G.R. & Pretty, J.N., *Unwelcome Harvest: Agriculture and Pollution*, Earthscan Publications, 1991
Windust, A., *Green Home Recycling*, Allscape, Victoria, 2000
Yeomans, P.A., *The City Forest: Keyline Plan for the Human Environment Revolution*, Keyline Publishing, Australia, 1971
Resources
Environmental Soil Testing Laboratory, Pymble, NSW
Soil Testing, Department of Agriculture, Rydalmere, NSW
Water Resources Commission

Chapter 7
Book
Suzuki, D., *Tree: A Biography*, Allen & Unwin, Sydney, 2001
Resources
On Borrowed Time, The Potter Farm Plan in Action, 1990, video, distributed by Trust For Nature, Melbourne
Schumacher, E.F., *On the Edge of the Forest*, video, Western Australia, 1977

Chapter 8
Books
Bubel, N., *The New Seed-Starters Handbook*, Rodale Press, Emmaus, USA, 1988
Fanton, M. & J., *Seedsavers' Handbook*, Seedsavers Network, Byron Bay, Australia, 1993 (www.seedsavers.net)
Morrow, R., *Family Seedsaving Book*, Wildfire Press, Blackheath, Australia
Wolverton, Dr B.C., *How to Grow Fresh Air: 50 Houseplants that Purify Your Home or Office*, Penguin Books, 2004 (www.wolvertonenvironmental.com)
Websites
www.organicgardening.com
www.permacult.com.au

Chapter 9
Refer to the books listed under the heading Important General References.

Chapter 10
Refer to the books listed under the heading Important General References.
Websites
www.frappr.com/permaculture
www.srd.org.au

Chapter 11
Books and magazines
Alexander, C., et al., *A Pattern Language*, Oxford University Press, New York, 1977
Alternative Technology Association (ATA), 'Green Power Buyer's Guide', *ReNew Magazine*, Oct–Dec 2005
Australian Government, *Your Home* (www.yourhome.gov.au)
Hollo, N., *Warm House: Cool House: Inspirational Designs for Low-energy Housing*, Choice Books, Sydney, 1995
Jenkins, J., *The Humanure Handbook*, 3rd ed., Jenkins Publishing, USA, 2005 (www.jenkinspublishing.com)
Pearson, D., *The Natural House Book*, Angus & Robertson, Sydney, 1990
Steen, B. & Baingridge, D., *The Strawbale House*, Green Books, 1994
Websites
Energy situation:
www.ata.org.au
www.basix.nsw.gov.au
www.ecologicalhomes.com.au
www.ecosource.com.au
www.greenbuilder.com
www.greenhouse.gov.au
www.greenpages.org
www.sustainabilitycentre.com.au
Solar photo-voltaic electricity:
www.ata.org.au
www.originenergy.com.au
www.solarenergy.org
Wind energy:
www.energybulletin.net
www.scoraigwind.com

Green power and solar-thermal hot water:
www.greenpower.gov.au

Chapter 12
Books
Cleveland, D.A. & Soleri, D., *Food from Dryland Gardens*, Centre for People, Food and Environment, Tucson, USA, 1991

Creasey, R., *The Complete Book of Edible Landscaping*, Sierra Club Books, San Francisco, 1982

Davies, J., *The Victorian Kitchen Garden*, BBC Books, London, 1991

Ellis, B.W. & Bradley, F.M. (eds), *Rodale's All-New Encyclopaedia of Organic Gardening*, Rodale Press, USA, 1990

Halweil, B., *Home Grown: The Case for Local Food in a Global Market*, Worldwatch Paper 163 (www.worldwatch.org/pubs/paper/163/)

Tudge, Colin, *So Shall We Reap*, Penguin Books, London, 2004

Websites
www.acfonline.org.au
www.echonet.org
www.edibleforestgardens.com
www.haryana-online.com
www.herbsarespecial.com.au
www.greenharvest.com.au

Chapter 13
Glowinski, L., *The Complete Book of Fruit Growing in Australia*, Thomas Lothian, Melbourne, 1991

Johns, V. & Stevenson, L., *Fruit for the Home Garden*, Angus & Robertson, Sydney, 1986

Chapter 14
Daghir, N.J. (ed), *Poultry Production in Hot Climates*, Faculty of Agricultural Sciences, United Arab Emirates University, April 1995

De Baracli-Levy, J., *Herbal Handbook for Farm and Stable*, Faber & Faber, London, 1986

French, J., *Jackie French's Chook Book*, Aird Books, Melbourne, 1993

Chapter 15
Books
Allen, A., *Growing Nuts in Australia*, Night Owl, 1986

Forbes, J.M., *Voluntary Food Intake and Diet Selection in Farm Animals*, University of Leeds, UK, 1995

Fukuoka, M., *Natural Way of Farming*, Japan Publications, Tokyo, 1985

Newell, P., *The Olive Grove*, Penguin Books, Ringwood, Vic., 2000

SEASAN (South East Asia Sustainable Agriculture Network), *Resource Book on Sustainable Agriculture for the Tropical Lowlands*, Bangkok, 1992

Solbrig, O. & D., *So Shall You Reap: Farming and Crops in Human Affair*, Shearwater Books, Island Press, USA, 1994

Website
www.bfa.com.au

Chapter 16
Books
Bootle, K.R., *Commercial Timbers of NSW and Their Uses*, Building Information Centre, Sydney

Douglas & Hart, *Forest Farming*, Watkins, London, 1980

Haddlington, P. & Johnston, J.A., *Australian Trees: A Guide to Their Care and Cure*, University of NSW Press, 1983

NSW Dept Agriculture, *Farm Trees*, Series 1–5, Trees on Farms Program, 1988

Oates, N., *Trees on Farms*, Goddard & Dobson, Victoria, 1988

Oates, N. & Clarke, B., *Trees for the Back Paddock*, Goddard & Dobson, Sydney, 1987

Wilson, E.O. (ed), Peter, F.M. (assoc. ed), *Biodiversity*, National Academy Press, Washington DC, 1988

Website
www.treesforhealth.org

Chapter 17
Books
Bradley, J., *Bringing Back the Bush*, Lansdowne Press, Australia, 1988

Bretchwold, R., *Wildlife in the Home Paddock*, Angus & Robertson, Sydney, 1985

Websites
www.conservationvolunteers.com.au
www.foe.org
www.greenpeace.org
www.menofthetrees.org.au
www.rainforest.alliance.org
www.ran.org

Resources
Australian Association of Bush Regenerators
Bushcare and Landcare groups
Greening Australia
Trees on Farms program, Department of Agriculture
TAFE courses

Chapter 18
Books
Cable, M. & French, F., *The Gobi Desert*, The Macmillan Company, New York, 1944

Mollison, B., *Arid Land Permaculture*, Tagari Publications, Australia, 1978

Chapter 19
Books
Fujixerox, *Down-to-Earth Office Care Guide* (www.officecare.info)

Heeks, A., *The Natural Advantage: An Organic Way to Grow Your Business*, Rodale Press, Emmaus, USA, 2001

Websites
www.corpwatch.org
www.energyrating.gov.au
www.energystar.gov.au
www.fujixerox.com.au
www.greenglobe.com
www.organicstyle.com

Resource
Green Building Council of Australia

Chapter 20
Books and magazines
Anon., *Disaster*, NSW State Emergency Services
Anderson, I. & Cross, M., 'World Faces Growing Quake Threat', *New Scientist*, 17 Dec 1988
Webster, J.K., 'Is Blowing on the Wind', *Town and Country Farmer*, Summer 1987
Website
www.bushfireinfo.com

Chapter 21
Books and magazines
Chapman, B., Penman, D. & Hicks, P., *Natural Pest Control*, Nelson, Melbourne, 1987
French, J., *The Organic Garden Doctor*, Angus & Robertson, Sydney, 1988
Hockings, F.D., *Friends and Foes of Australian Gardens*, A.H. & A.W. Reed, 1985
McMaugh, J., *What Garden Pest or Disease is That?* Lansdowne Press, Sydney, 1985
New, T.R., *Associations Between Insects and Plants*, University of New South Wales Press, 1988
Rogers, P., *Organic Pest Control*, Kangaroo Press, 1995
Sattaur, O., 'Pheromones Add a New Twist To Cotton Crop', *New Scientist*, 7 Jan. 1989
Websites
www.academicpest.com.au
www.neeminaustralia.com
Resources
Henry Doubleday Research Association
Organic Growers Association

Chapter 22
Books
Bretchwold, R., *Wildlife in the Home Paddock*, Angus & Robertson, Sydney, 1985
Clairvaux, P., *Why Big Fierce Animals are Rare*, Penguin, 1991
Clyne, D., *Best of Wildlife in the Suburbs*, Oxford University Press, Melbourne, 1993
Clyne, D., *Care of Urban Wildlife*, Taronga Park Zoo, 1996
Websites
Check the websites for the organisations listed in the resources.
Resources
Journals from the World Wildlife Fund, the Australian Conservation Foundation (*Habitat*), the CSIRO (*Ecos*), Sierra Club, World Watch Institute, Department of National Parks & Wildlife, Friends of the Earth, The Wilderness Society, Total Environment Centre, the International Union for Conservation of Nature
On Borrowed Time, The Potter Farm Plan in Action, 1990, video, distributed by Trust For Nature, Melbourne

Chapter 23
Books
Australian Institute of Agricultural Science, *The Threat of Weeds to Bushland*, Inkata Press, Sydney, 1976
Buchanan, R., *Bush Regeneration*, TAFE, Sydney, 1989
French, J., *Organic Control of Weeds*, Aird Books, Flemington, Australia, 1990

Resources
Society for Growing Australian Plants
Wildplant Rescue Service

Chapter 24
Books
Romanowski, N., *Fish and Farm Dams in Australia*, Melbourne, 1989
Sainty, G. & Jacobs, S., *Water Plants in Australia*, Australian Water Research Council
Website
www.amcs.org.au
Resources
Alternative Aquaculture Journal, Hawaii
Dept of Agriculture and Fisheries

Chapter 25
Books
Action for World Development, *New Ways with Old Money*, Education Kit, 2nd ed., 1991
Deniss, R., & Hamilton, C., *Affluenza: When too Much is Never Enough*, Allen & Unwin, Sydney, 2005
Hamilton, C., *Growth Festish*, Allen & Unwin, Sydney, 2002
Hawken, P., Lovins, A.B. & Hunter Lovins, L., *The Next Industrial Revolution*, Earthscan Publications, London, 1999
Roddick, A., *Take it Personally*, Harper Collins, Sydney, 2001
Websites
www.calvertgroup.com
www.communityinvest.org
www.eiris.org
www.ethicalmoney.org
www.jubilee2000uk.org
www.pcuonline.org

Chapter 26
Books
Andruss, V. et al., *Home: A Bioregional Reader*, New Society Publishers, Philadephia, USA, 1990
Mollison, B., *Bioregional Organisations*, Permaculture Institute, Tasmania
Websites
www.bioneers.org
www/fairtrade.org.uk
www.ilsr.org
www.oxfam.org.au
www.wdm.org.uk

Chapter 27
Books
Cartwright, R.M., *The Design of Urban Space*, Architectural Press, London, 1980
Giraudet, H., *The Gaia Atlas of Cities*, London, 1996
Ivanko, J. & Kivirist, L., *Rural Renaissance*, New Society Publishers, Canada, 2004
Websites
www.communitysolution.org
www.endofsuburbia.com
www.holmgren.com.au
www.responsibleshopper.org
www.sweatshops.org

Index